Guide for Plant Appraisal

10th Edition

Revised

International Society of Arboriculture
Atlanta, GA

ISBN: 978-1-943378-02-9

Copyright © June 2019 by International Society of Arboriculture.
All rights reserved. In accordance with the U.S. Copyright Act of 1976, the scanning, storing in any database, uploading, and electronic sharing of any part of this book without the permission of the publisher is unlawful piracy and theft of intellectual property. Unauthorized reproduction of copyrighted materials, in any format, can result in statutory penalties and legal prosecution. If you would like to use material from the book (other than for review purposes), prior written permission must be obtained by contacting ISA.

The ISA logo is a registered trade and service mark.

International Society of Arboriculture
270 Peachtree Street NW, Suite 1900
Atlanta, GA 30303
United States
+1 (678) 367-0981
www.isa-arbor.com
E-mail: isa@isa-arbor.com; permissions@isa-arbor.com

Editorial and Publishing Manager: Stephanie Ebersohl
Technical Editor: Colleen Mullen
Graphic Designers: Beatriz Pérez González and Amy Reiss
Illustrators: Troy Courson, Image Graphics Enterprises, Inc., and Beatriz Perez Gonález
Printed in the United States of America by Premier Print Group, Champaign, IL

10 9 8 7 6 5 4 3

1020-JV-1100

2130 Stella Court
Columbus, OH 43215, U.S.

9707 Key West Avenue, Suite 100
Rockville, MD 20850, U.S.

12500 Fair Lakes Circle, Suite 200
Fairfax, VA 22033, U.S.

SINCE 1948

312 Montgomery Street, Suite 208
Alexandria, VA 22314, U.S.

636 Eye Street NW
Washington, DC 20001, U.S.

270 Peachtree Street NW, Suite 1900
Atlanta, GA 30303, U.S.

136 Harvey Road, Suite 101
Londonderry, NH 03053, U.S.

Guide for Plant Appraisal, 10th Edition
Authored by representatives of the
Council of Tree & Landscape Appraisers

Richard F. Gooding Gooding's Nursery & Landscaping Sherrodsville, OH	**AmericanHort**
James R. Clark HortScience, Inc. Pleasanton, CA	**American Society of Consulting Arborists**
Pieter Severynen Pieter Severynen Associates Los Angeles, CA	**American Society of Landscape Architects**
Bret P. Vicary Sewall Forestry & Natural Resource Consulting Old Town, ME	**Association of Consulting Foresters of America, Inc.**
Mark Duntemann Duntemann Urban Forestry, LLC South Royalton, VT	**International Society of Arboriculture**
Len Burkhart The Davey Tree Expert Company San Diego, CA	**National Association of Landscape Professionals**
E. Thomas Smiley Bartlett Tree Research Laboratory Charlotte, NC	**Tree Care Industry Association**

Edited, published, and copyrighted by the
International Society of Arboriculture

Tenth Edition
June 2019
Revised October 2020

Contents

Disclaimer

This book is intended to be used for informational purposes and provides general information as a reference and procedural guide for plant appraisal, and it should not be considered a standard. This book is not intended to discourage appraiser choice. As reflected in the title, the *Guide for Plant Appraisal* is a resource for the plant appraiser. It is, however, neither an absolute nor a complete treatment of the subject of plant appraisal.

Appraisal approaches, processes, and methods are continually evolving, modifying, and progressing, as illustrated from each subsequent edition of the *Guide for Plant Appraisal*. While the authors, Council of Tree & Landscape Appraisers (CTLA) member organizations, and publisher believe that the information contained in this book is correct and complete to the best of their knowledge, no warranty is given or implied as to its accuracy. All recommendations are made without guarantee on the part of the publisher or CTLA member organizations.

In no event will the authors, CTLA member organizations, or publisher be liable for any loss or damage, including, without limitation, direct, indirect, or consequential loss or damage, or otherwise any loss or damage whatsoever, arising in contract, tort, or any other action, which may arise as a result from the use of or failure to use the information in the *Guide for Plant Appraisal*, and disclaim any liability in regard thereof. This text does not warrant nor is it intended to create any bias toward a specific treatment.

This text is published by the International Society of Arboriculture under agreement with the CTLA member organizations.

The Council of Tree & Landscape Appraisers members include AmericanHort, American Society of Consulting Arborists, American Society of Landscape Architects, National Association of Landscape Professionals, Association of Consulting Foresters of America, International Society of Arboriculture, and the Tree Care Industry Association.

Acknowledgments

The Council of Tree & Landscape Appraisers (CTLA) gratefully acknowledges the contributions of its previous members, including Jim Ingram, RCA #294 (ASCA); David Hucker, RCA #388 (TCIA); Tim Toland (ASLA); Logan Nelson (ASCA); Russ Carlson, RCA #354 (ISA); Scott Cullen, RCA #348 (ISA); Denice Britton, RCA #296 (ASCA); Robert Brudenell, RCA #417 (ASCA); and Lew Bloch, RCA #297 (NALP). Each had a significant influence on the content of this edition. The CTLA also appreciates the many years of leadership provided by Jim Ingram and David Hucker as chair.

Each of the CTLA's member organizations provided insight, comment, and feedback essential for the development of this edition. The CTLA is grateful to regional groups, standing plant appraisal committees, and appraisal workshop participants who continue to expand the knowledge base for the *Guide* and its readers. At ASCA, the A3G Committee was chaired by Jim Ingram, RCA #294. At ISA, the Plant Valuation and Appraisal Committee was chaired by Joe McNeil, RCA #299. At ACF, Lynn Wilson (executive director) and the Executive Committee were very supportive and encouraging.

Individuals who contributed significantly to this edition include Lori Ballard, Seminole, FL; Greg David, RCA #304, Muenster, TX; Steve Geist, RCA #340, Aurora, CO; Thomas Hanson, RCA #499, Kirkland, WA; Elizabeth Walker, Duvall, WA; James Komen, RCA #555, Glendora, CA; John Wickes, RCA #455, Spring Valley, NY; Greg McPherson, Davis, CA; Greg Carbone, Boston, MA; Ed and Lee Ann Steigerwaldt, ACF, Tomahawk, WI; John Goodfellow, Redmond, WA; Jud Scott, RCA #392, Carmel, IN; Lee Gilman, Amherst, MA; Randy Miller, CN Utility Consulting, Salt Lake City, UT; Dave Nowak, USDA Forest Service, Syracuse, NY; Dennis Yniguez, RCA #362, Berkeley, CA; and Barri Bonapart, attorney at law, Sausalito, CA.

Dr. Matt Ritter, California Polytechnic University, San Luis Obispo, provided excellent advice on botanical nomenclature and taxonomy.

Nelda Matheny, RCA #243, Pleasanton, CA, prepared several illustrations as well as a detailed review of the text. Thanks also to Dr. Timothy Toland for contributing to the illustrations.

The CTLA also thanks the ACF Board of Directors for their support.

The CTLA wishes to acknowledge the expertise, patience, and encouragement of Caitlyn Pollihan, Kevin Martlage, Marni Basic, Stephanie Ebersohl, and Amy Theobald of the International Society of Arboriculture. All provided a great deal of support to the CTLA and its mission.

The CTLA member representatives gratefully acknowledge the contributions of the executives at the council's seven green-industry organizations and their boards of directors: Bob Dolibois and Ken Fisher at AmericanHort; Lynn Wilson at ACF; Nancy Somerville at ASLA; Beth Palys at ASCA; Caitlyn Pollihan and Jim Skiera at ISA; Sabeena Hickman and James Leipold at NALP; and Mark Garvin at TCIA. Without their encouragement and support, the *Guide* would not have been possible.

Preface

It is now 60 years since the publication of the first edition of the *Guide for Plant Appraisal*. In preparing the tenth edition, the overarching objective of the CTLA was to provide the appraiser with a systematic process for defining the appraisal problem, identifying the appropriate appraisal approach(es), and developing a credible conclusion. In meeting this objective, the CTLA was guided by three goals.

The first goal was to build on and strengthen the approaches and methods presented in previous editions: cost, income, and sales comparison (formerly called *market*). The strengths and weaknesses of each plant appraisal approach and method are described. New research relevant to tree appraisal was incorporated, particularly that which related to the contribution trees make to real estate market value and the environmental and ecological benefits they provide.

The second goal was to lay a foundation for the process of plant appraisal. While previous editions of the *Guide* focused largely on methods, this edition provides detailed discussion of the concepts and processes associated with plant appraisal, and it stresses that the key step in the appraisal process is identifying the appraisal problem. Furthermore, this edition emphasizes differences among assignment results, particularly between cost and value estimates.

The third goal was to align the concepts and terminology of plant appraisal with those employed in the general practice of appraisal. This change will augment the plant appraiser's vocabulary so that they may understand the language used by other appraisers. Readers will find this most apparent in the cost approach, where the terms *reproduction* and *functional replacement* have been added and the depreciation factors revised.

The CTLA views this shift in terminology as similar to the International Society of Arboriculture's (ISA) effort to align tree risk assessment terminology with the general practice of risk assessment. While plant appraisers face unique circumstances, the ideas and concepts employed are the same as for general appraisal. To this end, the tenth edition modifies general terminology to better fit plant appraisal.

This edition also provides a foundation for understanding concepts related to highest and best use, contributory value, the principles of diminishing returns and balance, and others fundamental to economic theory and market behavior. Within this context, the unique aspects of plant appraisal are identified. Having acquired this foundation of understanding and application, plant appraisal should enjoy increased credibility within the larger appraisal community.

New to this edition are additional appendices and a glossary. Terms listed in the glossary are formatted in **bold** the first time they appear.

CHAPTER 1

Introduction

CHAPTER OUTLINE

Overview

Trees and landscapes have value. The appraisal process is used to identify a cost or value associated with plants and landscape features. Common reasons for plant appraisals include inventory, tree preservation, insurance, casualty loss, income, accounting, tax, finance, and litigation purposes. A plant appraiser may deal with a broad spectrum of plant valuations, from an individual tree to a wooded residential or recreation area, an industrial park, or an entire city.

The benefits provided by plants in urban, suburban, and rural settings are numerous (Figure 1.1). Over the past two decades, research has demonstrated the contribution that trees and landscapes make to real estate property value, as well as the beneficial role plants can play in conserving energy, removing atmospheric contaminants, moderating stormwater runoff, sequestering carbon, improving physical and mental aspects of human health, and increasing social capital (Jonnes 2016).

History

In 1916 at what is now the University of Massachusetts Amherst, Dr. George E. Stone introduced a formula for placing a monetary value on "shade and ornamental trees." Stone's formula considered tree size, species, and condition.

In 1938, Dr. Ephraim P. Felt, director of the Bartlett Tree Research Laboratories, and Orville Spicer, president of the F.A. Bartlett Tree Expert Company, revised Dr. Stone's formula (Table 1.1). The Felt-Spicer formula expanded Stone's work to include location and residential land values. A tree's initial value was based on a cross section of its trunk measured 4.5 feet (1.37 m) above the ground. This was adjusted by the location, condition, species, and land value, each expressed as a percentage, to obtain an appraised value. Considerable judgment was necessary as few criteria were provided with which to evaluate each of the rating factors.

Figure 1.1 Trees and other landscape plants provide a wide range of benefits.

The first edition of the *Guide* was *Shade Tree Evaluation*, published in 1957 through the joint efforts of the National Shade Tree Conference (now International Society of Arboriculture) and the National Arborist Association (now Tree Care Industry Association). The formula found in the first edition was similar to the Felt-Spicer formula except that the real estate and location ratings were dropped.

The *Guide* was revised in 1965 and 1970. Although not titled as such, they served as the second and third editions of the *Guide*, respectively. The fourth edition was published in 1975 with the title *A Guide to the Professional Evaluation of Landscape Trees, Specimen Shrubs, and Evergreens.*

Also in 1975, the International Society of Arboriculture and the National Arborist Association were joined by the American Society of Consulting Arborists (ASCA), the American Association of Nurserymen (now AmericanHort), and the Associated Landscape Contractors of America (now National Association of Landscape Professionals) to form the Council of Tree and Landscape Appraisers (CTLA). The CTLA published the fifth edition, titled *Guide for Establishing Values of Trees and Other Plants*, in 1979. The sixth edition was published in 1983, and a companion publication, the *Manual for Plant Appraisers*, was released in 1986. The seventh edition, *Valuation of Landscape Trees, Shrubs, and Other Plants: A Guide to the Methods and Procedures for Appraising Amenity Plants*, was published in 1988.

In 1996, the Association of Consulting Foresters joined the CTLA, followed by the American Society of Landscape Architects in 1997.

Prior to the eighth edition, the method of establishing unit cost for trees too large to be replaced (dollars per square inch of trunk cross-sectional area) was defined by the *Guide*. In 1957, unit cost was identified as \$5.00 per square inch (\$0.78 per cm^2). The unit cost increased with each edition to \$27.00 per square inch (\$4.19 per cm^2) by the seventh edition (1988). Note that throughout this *Guide*, all prices are in USD.

The first four editions described the formula method of appraisal for trees larger than replaceable size. A replacement method for small trees was added in the fifth edition. The delineation between replacement and formula was not well defined. For example, in the seventh edition, a replacement tree with a diameter of 9

Table 1.1 Highlights of plant appraisal in the United States and Canada. For a more detailed treatment, see Cullen (2005 and 2014).

1916	Stone (University of Massachusetts Amherst) prepares a formula for placing monetary values on "shade and ornamental trees"
1957	*Shade Tree Evaluation* (First edition)
1965	*Shade Tree Evaluation*, Revision I (Second edition)
1970	*Shade Tree Evaluation*, Revision II (Third edition)
1975	*A Guide to the Professional Evaluation of Landscape Trees, Specimen Shrubs, and Evergreens*, Revision III (Fourth edition)
1975	Council of Tree and Landscape Appraisers (CTLA) officially formed, composed of the International Society of Arboriculture, the National Arborist Association, the American Society of Consulting Arborists, and the American Association of Nurserymen (now AmericanHort)
1979	*Guide for Establishing Values of Trees and Other Plants*, Revision IV (Fifth edition)
1983	*Guide for Establishing Values of Trees and Other Plants* (Sixth edition)
1986	*Manual for Plant Appraisers* (First edition), published by the CTLA
1988	*Valuation of Landscape Trees, Shrubs, and Other Plants: A Guide to the Methods and Procedures for Appraising Amenity Plants* (Seventh edition)
1992	*Guide for Plant Appraisal* (Eighth edition)
2000	*Guide for Plant Appraisal* (Ninth edition)

inches (23 cm) was said to have a value of $44.00 per square inch, and the formula tree was said to be worth $27.00 per square inch.

The eighth edition, now called the *Guide for Plant Appraisal*, based the formula value of large trees on the replacement cost plus the increase in value due to the increase in tree size above that of the replacement tree. It recommended forming Regional Plant Appraisal Committees in order to determine what information was needed to appraise plants in localized geographic regions. Such information included the size of the largest transplantable tree, its cost, the cost-per-unit trunk area, and a species rating list.

A further refinement of the *Guide* formula, for the ninth edition in 2000, addressed the rapid increase in trunk area and, therefore, value of trees larger than 30 inches (75 cm) in diameter. Previously, trees over 30 inches could lead to exceptionally high and unreasonable appraised values. The *Guide* allowed an appraiser to adjust trunk area for those trees larger than 30 inches in diameter.

In preparing the ninth edition of the *Guide*, the CTLA identified three approaches to value: cost, income, and market. Aside from expanding upon the various methods that range from cost of repair and replacement cost to trunk formula and cost of cure, the CTLA compiled information on professional considerations and responsibilities and plant appraisal within easements and rights-of-way.

For a more complete discussion of the *Guide*'s history and development, see Cullen (2005).

Plant Appraisal and Property Value

With the exception of Felt (1938), early editions of the *Guide* did not make a connection between tree value and property value. Prior to the eighth edition (1992), plants were acknowledged mainly for their beauty and shade. Estimating value relied on developing a reproduction or replacement cost.

The eighth and ninth editions of the *Guide* developed the connection between plant appraisal and property value. The ninth edition noted that, "In the plant appraisal process, the value of individual plants and the whole landscape should be reasonably and closely dependent upon the land they occupy" (p. 10). Both the eighth and ninth editions emphasized that "appraised tree value should be 'reasonable.'" In this context, *reasonable* was described as the contribution of trees and landscape to real estate market value. This is a foundational aspect of establishing plant value.

Courts have recognized the disparity that may occur between market contribution and reproduction cost. While market contribution is the primary method for calculating damages, a strict adherence to market contribution may not necessarily make the tree owner whole. Courts have broad powers to interpret the facts presented, the context of the specific case, and case history and to make judgments accordingly. The subject of real estate market value regularly appears in plant appraisal litigation, and plant appraisers should be aware of discussion points that occur as a result.

Professionals within the green-industry are uniquely aware of property owners' appreciation of, and actual monetary expenditures toward, obtaining, installing, and maintaining trees and landscape plantings. Cost, however, does not equal value. The loss of a specimen tree may or may not result in a decrease in property value, even though the costs of clean-up and reproduction could be substantial. The connection between cost and value is discussed in Chapter 2.

A Guide Versus a Standard

As reflected in its title, the *Guide for Plant Appraisal* is a resource for the plant appraiser. It describes and summarizes concepts, approaches, and methods that may assist the plant appraiser in developing an opinion of value. In many ways, the *Guide* functions as a summary of best practices in plant appraisal. It is, however, neither an absolute nor a complete treatment of the subject of plant appraisal. Plant appraisers should be prepared to defend, explain, and support their work.

Industry **standards** are referenced throughout the *Guide*. These include both national publications (e.g., those produced by the American National Standards Institute [ANSI]) and state or regional publications (e.g., *Florida Grades and Standards for Nursery Stock*). For professionals in the green-industry, the term *standard* normally implies documents published by ANSI. ANSI standards focus on defining practice with *should* statements, which are advisory recommendations, and *shall* statements, which are mandatory requirements. They are prescriptive in nature.

The *Guide* is not an ANSI standard, even though it is produced through a consensus-driven process. The *Guide* is far less prescriptive and gives the plant appraiser greater flexibility than do ANSI standards.

APPRAISAL LICENSING

If trees and landscapes are real estate, and appraisers must be licensed by individual states, should plant appraisers also be licensed?

At the time of publication of the tenth edition, the CTLA was not aware of any instance in which a state real estate licensing board successfully challenged an individual's right to appraise plants on the basis that the individual did not possess an appraisal license or certification. The CTLA does not support a position that all plant appraisers should be licensed as real estate appraisers are.

Plant appraisers possess unique knowledge about the growth and development of trees and other plants; their use and function in the landscape; and the costs to obtain, install, and maintain them. A court of law may restrict the plant appraiser from opining on a property's market value but allow the plant appraiser to testify to the contribution of trees and landscape to that value and then consider the weight of the testimony in light of the appraiser's qualifications.

Plant appraisers should be aware that the term *standard* has a variety of meanings. For example, the International Organization for Standardization (ISO) describes a standard as a "document containing practical information and best practice; and often an agreed way of doing something or a solution to a global problem" (International Organization for Standardization 2004). ASTM International describes a standard as a "test method, manual, guide, practice, classification or terminology document" (www.astm.org). ANSI uses the ISO definition for a standard. Professionals outside the green-industry could define the term *standard* in a way other than ANSI.

The ISO describes a *de facto* standard as one that has been accepted by the marketplace. The *Guide* has a long history of acceptance by plant appraisers, their clients, and the courts, and the vast majority of plant appraisers rely on the *Guide* as their primary source of information.

In summary, the *Guide* is an important resource and reference that provides flexibility in the development of plant appraisals. It is a consensus document prepared and approved by the CTLA's member organizations. It is not designed as an ANSI Standard and should not be regarded as such. Because the *Guide* is not an ANSI Standard, plant appraisers should apply their craft and seek approaches appropriate to their specific assignment. Plant appraisers should always be prepared to defend, explain, and support their work.

Using the *Guide for Plant Appraisal*

This edition of the *Guide* is organized to guide the appraiser through the plant appraisal process. Chapter 2, "Core Concepts," introduces principles and ideas important to plant appraisal. Chapter 3, "The Appraisal Process," provides an overview of the process, including carrying out the initial contact, defining the assignment, gathering pertinent data, deciding on an appropriate approach, producing a result, and reporting.

This edition of the *Guide* describes three approaches to appraisal: cost, sales comparison, and income. Within each approach are one or more methods. In the cost approach (Chapter 5), there are three methods: repair, reproduction, and functional replacement. In the income approach (Chapter 6), the methods are either direct or related to benefits. The sales comparison approach (Chapter 7) focuses on one method: component analysis.

USPAP AND PLANT APPRAISAL

The Uniform Standards of Professional Appraisal Practice (USPAP) is recognized in the United States as a standard of appraisal practice, providing definitions, rules, and standards (The Appraisal Foundation 2016). The Appraisal Foundation oversees USPAP through the Appraisal Standards Board. Like the green-industry's ANSI standards, USPAP is updated and revised by a group of practicing professionals.

USPAP identifies two key ethical standards (The Appraisal Foundation 2016). First, an appraiser must promote and preserve the public trust inherent in appraisal practice by observing the highest standards of professional ethics. Second, an appraiser must perform assignments with impartiality, objectivity, and independence and without accommodation of personal interest. This includes avoiding the following: performing with bias, advocating the interest of any party, accepting an assignment that includes a predetermined conclusion, and communicating misleading information.

Plant appraisers using the *Guide* should consider USPAP a useful resource. Being knowledgeable about USPAP can strengthen and enhance the quality and credibility of plant appraisals.

CHAPTER 2

Core Concepts

CHAPTER OUTLINE

Overview

The purpose of this chapter is to introduce common appraisal terms and concepts. It highlights the core concepts of appraisal, including value, cost, and the economic principles that provide the foundation for plant and landscape valuation. The procedures for appraising plants and landscapes can vary widely depending on the facts and circumstances of the assignment. Understanding common appraisal terms, concepts, and principles will prepare the appraiser to apply the most appropriate valuation procedures to landscape assets.

Appraisal Defined

The term **appraisal** has been defined many different ways by trade organizations, government agencies, and the courts. The Appraisal Institute defines appraisal as "the act or process of developing an opinion of value." **Valuation** is a synonym for appraisal and therefore may be used interchangeably with it (Appraisal Institute 2015).

The end product of the appraisal process is an **assignment result**, "an appraiser's opinions and conclusions developed specific to an assignment" (The Appraisal Foundation 2018). The assignment result is an opinion or estimate, not a fact.[1] The appraisal problem, terms of the assignment, and nature of the appraiser's work will

[1]Some plant appraisers use **appraised value** to refer to the assignment result, though it calls into question what type of cost or value it represents.

determine whether the assignment result should be a cost or a value. The appraisal report should provide a definition of the type of cost or value being estimated (e.g., reproduction cost, repair cost, market value) and a citation for the definition.

Plant appraisers specialize in plants, landscaping, and associated services, including the production of **crops**. **Plant appraisal** is the act or process of formulating an opinion of a defined cost or value for plants, landscape elements, or services.

Price, Cost, and Value

In casual conversation, price, cost, and value are often used interchangeably. However, it is important for professionals to distinguish among them (Figure 2.1). **Price** is the amount of money asked, offered, or paid for **property** or services. Once stated, it is a fact, not an estimate (Appraisal Institute 2015). It also can be the amount of money that a seller is asking for property or services. Price may or may not relate to value or cost. To the appraiser, price is an observable data point, not something that the appraiser estimates. For instance, one can collect price data from a nursery or a listing price from a real estate broker.

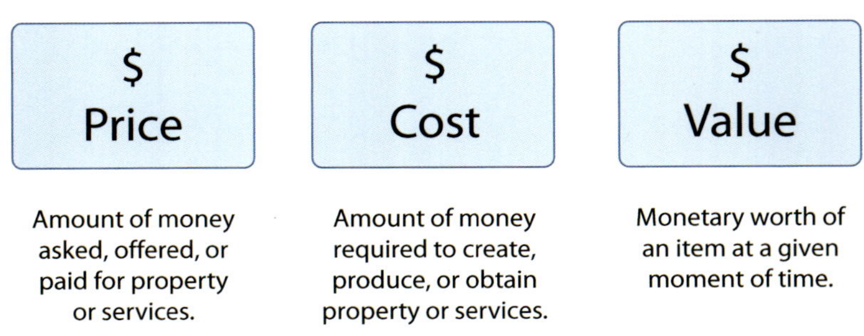

Figure 2.1 Comparison of price, cost, and value.

Cost is the amount of money required to create, produce, or obtain a property or service (Appraisal Institute 2015). The term is used either as a historic fact or as an estimate of current, future, or historic **reproduction cost** or **functional replacement cost** (Appraisal Institute 2015). Cost may include profit margin.

Value is the monetary worth of an item at a given time (Appraisal Institute 2015). It is created by the expectation of future **benefits** in the minds of sellers, buyers, and users of assets. Appraisers may use both cost and price information to develop estimates of value.

As noted in the introduction, cost does not equal value. The cost of an item may be more or less than its value, and in some cases the cost of an item may have little or no relationship to its value.

Cost Concepts

Cost estimates are the primary focus of valuation for many plant appraisers. The starting point for developing a cost estimate is determining the cost to purchase a new plant or landscape item. There are three fundamental ways to apply the cost approach (Figure 2.2). One is to develop a reproduction cost, or the cost to produce an exact (or nearly exact) replica of a landscape item. The second is to develop a functional replacement cost, or the cost to replace a landscape item with an item having equivalent **utility**. The third is an estimate of the cost to repair an item. Cost estimates may also be used as inputs for estimating expenditures necessary for repairing or mitigating deterioration caused by decay, wear and tear, or partial destruction.

Depreciation

Depreciation is the monetary expression of suboptimum factors. Appraisers use depreciation to account for the differences between the cost of the new or ideal item and the item being appraised, which typically has some lower level of quality due to its less than ideal features, its placement, or the site that it occupies.

There are three basic components of depreciation (Appraisal Institute 2013a): physical deterioration, functional obsolescence, and external obsolescence. For plant appraisal, **physical deterioration** is referred to as condition (health, structure, and form). Functional obsolescence is termed **functional limitations** and describes features that restrict or constrain growth or function due to poor placement, excessive size, or quantity.

External obsolescence is termed **external limitations**, and includes those legal, biological, and environmental conditions external to the property.

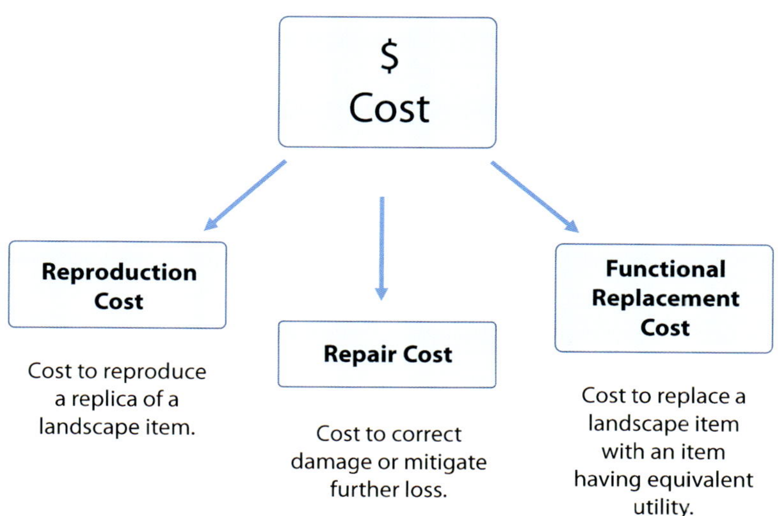

Figure 2.2 Application of the cost approach reflects reproduction, repair, or functional replacement.

A cost estimate may be the desired assignment result, either with or without depreciation. Depreciation is a loss of value from any cause, typically caused by either physical, economic, or external factors. In either case, the appraiser focuses on cost, with no particular regard to **willingness to pay** (discussed on page 10). In other cases, the appraiser may use a cost estimate as the basis for estimating value, in which case depreciation should reflect the difference between cost and market value (Appraisal Institute 2015). The type and objective of the assignment will determine which methods are appropriate.

Value Concepts

Value is the monetary relationship between an item and those who buy, sell, or use it (The Appraisal Foundation 2016). Anticipation of future benefits creates utility, and utility creates value. In appraisal, value is not a fact but an opinion of the worth of a property at a given time and in accordance with a specific definition of value. For economic value to exist, there must be not only **demand** but also some degree of **scarcity**, or limited supply. Where market value is sought, the supply and demand balance in the marketplace ultimately determines an item's value. Appraisers estimate value, whereas buyers and sellers determine value.

The appraiser defines the type of value to be appraised at the outset of the assignment. Most value-based appraisal problems require estimating what someone is willing to pay to obtain the benefits provided by the asset. Evidence of willingness to pay can be derived from previous purchases. Examples of value include **market value**, **investment value**, and **liquidation value**. In plant appraisal, landscape plants have market value because they are **real estate** and are generally traded as part of a larger real estate property. The present value of net income produced by a nursery, woodlot, or vineyard may represent market value; when using client-specific inputs, it may represent investment value. Trees killed by fire or insects may have some liquidation value.

Other types of value are non-transaction-based and should be analyzed in terms of inferred supply and demand. Examples of non-transaction-based value include **ecosystem value**, **existence value**, and **public interest value**. Ecosystem values may include benefits associated with carbon sequestration, temperature control, biodiversity, and wildlife habitat. The value of some **ecosystem services** can be measured using willingness-to-pay models or inferred from tree management software applications, like **i-Tree Eco**, which estimate the value of environmental and ecological functions. The value of landscape plants associated with public property (such as parks and streets) and institutions are inferred because these properties are not transacted in the marketplace. Historic trees may have public interest value.

The most common type of value appraisal is market value, which may be defined as "an opinion of value that presumes the transfer of a property, as of a certain date, under specific conditions set forth in the definition of the term identified by the appraiser" (The Appraisal Foundation 2016). It is also known as **fair market value.** The various definitions of market value share a common theme: the presumption of an open-market sale free of abnormal influences, where seller and buyer are knowledgeable and act in their own interests, and cash is exchanged on a specified date.

Willingness to Pay

Value estimates are generally based on willingness to pay (WTP), or the notion that an item's value can be demonstrated by how much one would be willing to pay for it. Willingness to pay is tied to the **principle of substitution** (discussed on page 13). Market value and cost estimates are both based on the actions of buyers and sellers. Transaction data reflect supply and demand and provide the basis for valuation.

Public parks, trails, and other green spaces are generally located on properties that are not bought and sold. In such situations, **non-market valuation** methods may be used to estimate the values of such resources. These **non-market values** are not based on exchange or actual transaction, but they can be demonstrated by willingness to spend time and money or estimated through willingness to support public funding by paying higher taxes. Such inferred price data provides a basis for estimating willingness to pay.

Trees and other landscape items have tangible and intangible value to their owners. Tangible benefits include energy savings and fruit, some of which can be quantified. Intangible values include aesthetics, sentimentalism, and privacy. These values may be more or less independent of the land that the items occupy, and may be independent of the true market value of the landscape items themselves. Intangible benefits can be extremely difficult to objectively measure and vary from one owner to another.

There is often a disconnect between the owner's perception of a tree's value, its cost, and its market value. For example, a specimen shade tree producing various tangible and intangible benefits may be worth $10,000 to its owner (e.g., the owner might state a willingness to pay $10,000 to avoid losing it). The tree may have a reproduction cost of $25,000, but it may add only $1,000 to the overall property value.

Intangible values unique to the landowner or client, such as aesthetics or sentimentalism, are likely to be personal and subjective. Such personal values (also known as value for reasons personal) may not reflect more general perceptions of utility or be supported by transactions in the marketplace. The appraisal process, however, should be objective and impersonal. The appraiser should apply reasonable industry standards to the appraisal process, not subjective personal opinions offered by others. Where market value is sought, transactional behavior should ultimately provide the foundation for estimating the amount that the plant or landscape item contributes to overall property value.

Highest and Best Use

Market value is estimated based on an asset's **highest and best use** (**HBU**). It is defined as "the reasonably probable use of property that results in the highest value." The four criteria that the highest and best use must meet are: (1) legal permissibility, (2) physical possibility, (3) financial feasibility, and (4) maximum productivity (Appraisal Institute 2013a).

The term *probable* is widely interpreted as "reasonably probable within the reasonably foreseeable future." When appraising the market value of plants or landscape services, plant appraisers should consider the current use and potential utility of the overall property *before* deciding which valuation approach and method to use. HBU may be described using land-use terms such as residential, commercial, industrial, real estate, agricultural, timber, and recreation. Whether HBU should be considered is a function of the appraisal problem.

In most cases, a property's HBU will be obvious from the current use of the property and the neighborhood characteristics. In other cases, the plant appraiser may need to consult with local professionals (e.g., real estate broker, real estate appraiser). Such situations could involve a forested property in transition to commercial use with the expectation of future development or a single tree on an undeveloped vacant lot within a developed residential subdivision.

Cost Estimates Versus Market Value Estimates

Cost estimates are based on production costs. In plant appraisal, this includes acquisition of plant and landscaping materials, plus labor and costs associated with transportation, installation, and aftercare. When these plants and materials are installed, personal property (e.g., plant specimens, brick and mortar) is converted to real estate (e.g., installed trees and patios).

Value is based on willingness to pay. Where market value is sought, evidence of WTP derives from transactions. In contrast, value estimates focus on the contribution of trees and landscape to overall property value.

Types of Property

There are two general types of property: real estate and personal property. Owners of both types of property possess sets of legal rights in the asset. For example, where the appraiser is asked to estimate the cost to replace a tree or to restore a site to some previous condition or utility, it may be appropriate to use cost methods without particular regard for market value or transaction behavior. Where an assignment calls for estimating market value of plants and other landscape items, the appraiser estimates their value in the context of overall property value, or what they contribute to the value of the whole.

Real Estate

Real estate is the "physical land and appurtenances attached to the land" (Appraisal Institute 2013a). The legal definition of real estate includes the following tangible items (Appraisal Institute 2013a):

1. Land
2. All things that are a natural part of the land, such as trees and minerals
3. All things that are attached to the land by people, such as buildings and site improvements (e.g., fences or landscaping)

Real estate appraisal seeks to estimate the cost or value of the above three items. There are two distinguishing features of real estate:

1. Its utility and value are a function of its location.
2. It is intended to be used in its current location, not moved to some other more permanent site.

Where trees, shrubs, and other landscape features exist in their final location, they constitute real estate rather than personal property. This is true for both natural plant communities and designed landscapes. For example, consider the following from the United States federal tax code:

> *[Timber is] defined by the Internal Revenue Code as the wood in standing trees that is to be recovered when the trees are cut and processed... Once trees are cut, they cease to be timber for income tax purposes. The simple act of cutting standing trees converts timber from real property to wood products and personal property.* (Passewitz)

Where a home owner pursues a deduction for loss or damage to ornamental or shade trees, the Internal Revenue Service (IRS) ordinarily recognizes the depreciation in the value of the entire property (land and improvements) as the appropriate measure of the value of the loss (The West Group 2005). The treatment of plants and other site improvements as real estate is by no means limited to public agencies and courts. It is a fundamental tenet of **real property** appraisal and taxation. So where components of a property are damaged, it is common for damages to be measured in terms of loss in market value.

These basic real estate concepts have profound implications for the plant and landscape appraiser. For example, where the appraiser is asked to estimate the cost of replacing a tree or of restoring a site to some previous condition or utility, it may be appropriate to use cost methods without particular regard for market value or transaction behavior. Where an assignment calls for estimating market value of plants and other landscape items, the appraiser estimates their value in the context of overall property value, or what they contribute to the value of the whole.

Personal Property

Personal property includes all tangible property that is not classified as real estate, as well as intangible property (e.g., **goodwill**, reputation, customer lists) (Appraisal Institute 2015). Some distinguishing features of personal property are as follows:

1. Utility and value are not generally tied to location.
2. Use occurs across a variety of locations.
3. It is moveable without damage to itself or the real estate (Appraisal Institute 2013a).

Nursery stock is personal property because it is part of a business stock in trade (Figure 2.3). It is intended to be moved and installed in some other permanent location. Outplanting nursery stock to a field for subsequent removal from the site still means the nursery stock is personal property. It becomes real estate once it has been installed in a permanent location. Its value thus becomes related to its location, though the cost to replace or restore it may not be. If it is subsequently transplanted to another location, it can actually become personal property again and then convert back to real estate once installed.

Figure 2.3 Progression of trees from personal property to real estate.

A standing tree in the forest is real estate, but it becomes personal property once it is severed from the stump. Even Christmas trees are real estate once planted in the grower's field; they become personal property when they are cut. Trees planted in someone's yard are real estate. Containerized plants, such as bonsai, are personal property, since one would expect an owner who vacates a property to remove such specimens and carry them away.

Economic Principles

Several economic principles have a direct bearing on the appraiser's analysis (Figure 2.4). An understanding of these principles provides a comprehensive understanding of concepts that may be of use to a plant appraiser.

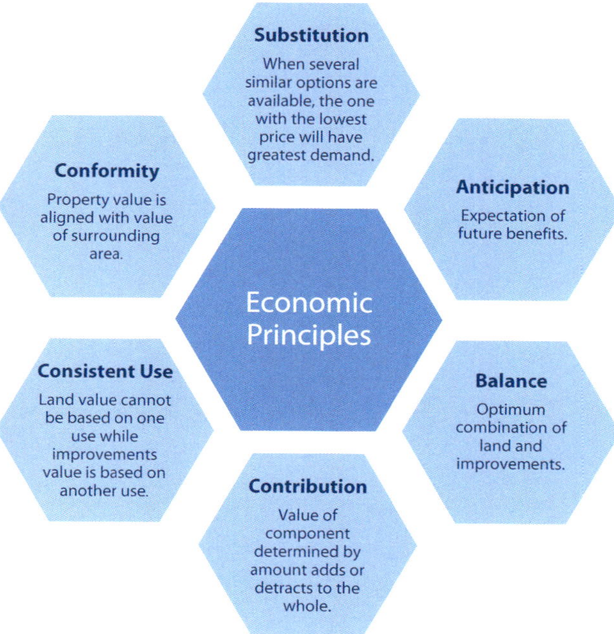

Figure 2.4 Economic principles important to plant appraisal.

Principle of Substitution

The **principle of substitution** states that when several similar or commensurate commodities, goods, or services are available, the one with the lowest price will attract the greatest demand and widest distribution (Appraisal Institute 2015). This is the primary principle upon which the cost and sales comparison approaches rest.

In the context of estimating market value, the principle of substitution may be restated by asserting that a rational buyer will not pay more for a particular property than the cost to obtain a substitute property having similar utility. This principle is helpful to the appraiser who needs to decide whether reproduction cost or functional replacement cost should be used, and for establishing whether cost and value estimates are reasonable in relation to consumer behavior.

For example, a hedge of exotic species that was used for screening is destroyed. It is possible to reproduce with the same species for $10,000 or replace it with a hedge of local species for $2,000 without sacrificing any utility. The principle of substitution suggests that the $2,000 figure may be a more reasonable valuation depending upon the circumstances (e.g., where reproduction cost is not required).

Another example is where a 15-year-old forest crop has been damaged and is to be replaced. Customary practice is to plant containerized seedlings rather than install 15-year-old trees. After adjusting for the time and costs to grow seedlings for 15 years, the cost to restore the forest is far less than the cost to install large trees, and it is consistent with normal forestry practices. The same process may apply to damages to ornamental trees and shrubs.

Principle of Anticipation

The **principle of anticipation** states that value is created by the expectation of future benefits (Appraisal Institute 2015). This is the primary principle upon which the income approach rests. For example, investors evaluate vineyards, orchards, Christmas tree farms, and timber properties in terms of the potential income that they can provide. Here, market value is reflected by the present value of future net cash flow.

One of the unique aspects of plant appraisal is that the asset increases in size over time. Because environmental and ecological benefits increase as a tree grows larger, it is reasonable to anticipate that the value of those benefits will increase as well. The current and anticipated benefits can be modeled and their value estimated using tree management software such as i-Tree Eco.

With market value, the principle of anticipation also applies because value is based on the expectation of future benefits and the demand for a given property. In contrast, past or anticipated costs may, in certain cases, have little relevance to anticipated benefits.

Principle of Balance

The **principle of balance** applies to the relationship among various property components such as land, trees, and buildings. It rests on the fact that property value is maximized when contrasting, opposing, or interacting components are in a state of equilibrium. Economic balance exists where there is an optimum combination of land and improvements, i.e., when no additional benefit is achieved by adding another unit of capital (investment in improvements) (Appraisal Institute 2013a).

For example, consider a residence with a house and landscaping that are typical for its neighborhood. One could double the investment in house or landscaping, but it is unlikely that this would result in a commensurate increase in property value. The installation of $100,000 in landscape improvements would likely add more value to an upscale estate property than to a small lot in a modest subdivision. These examples reflect the **law of diminishing returns**.

The law of diminishing returns operates at the residential landscape level. Research has demonstrated that people prefer unimproved lots with a few trees over lots with either no or a high number of trees (Payne and Strom 1975). Once the "optimum number" of trees is in place, adding another tree may have little or no positive impact on the value of the property and may result in a decrease in property value.

Excessive landscaping may even detract from property value. The market value of an "excess" tree may be much less than its **replacement cost**, or even zero. This illustrates functional limitations due to a **superadequacy** (see Chapter 5), or an excess in the capacity or quality of a structure or structural component which does not add value or functional utility to an object or property.

Principle of Contribution

The **principle of contribution** holds that the value of a particular component of a property is measured in terms of its contribution to the value of the whole property, or the amount by which its absence would detract from the value of the whole (Appraisal Institute 2013a, 2015). This amount is called **contributory value**.

Trees and other plants are real estate, and rarely trade in the absence of the sale of the entire property of which they are a part. Therefore, their market value is measured in terms of their contribution to the whole.

A landscape item's contributory value may be higher than, equal to, or less than its cost. For example, a specimen tree may cost $10,000 to install. This does not mean that a typical buyer would pay either $10,000 more for the property after it is installed or $10,000 less for the property if it were removed. The same principle applies to special landscape amenities, such as swimming pools and tennis courts.

A landscape item might even detract from property value if it is an invasive species, blocks views, produces excessive litter, or needs to be removed due to poor health or unsatisfactory location.

For many plant appraisal projects, particularly where the assignment calls for estimating the cost to reproduce property or restore it to a previous condition, contributory value may not be particularly relevant.

Principle of Consistent Use

The **principle of consistent use** holds that land cannot be valued based on one use while improvements (e.g., landscaping and structures) on the land are valued based on another use (Appraisal Institute 2013a). This principle applies primarily to market value analysis.

For example, a parcel of natural woodland lies between two high-value residential properties. The area is zoned residential. The wooded parcel would be considered an interim use. The typical buyer of the parcel would clear all or some of the land, build a house, and install a landscape typical for the neighborhood. If, prior to the development, a neighbor were to damage some of the natural trees and ground cover, it may be incorrect to conclude that the market value of damaged plants is commensurate with the cost to replace them, particularly if it is likely that the plants would be removed in the development process.

This example reflects a relatively common situation where natural forest trees occupy property that is to be converted into a residential subdivision. Whether the existing trees contribute to the property's market value depends on what the typical buyer would do with them. It is possible that certain trees would be retained during development for natural screening or shade. Other trees would be removed if they impede development, obstruct views, lack landscape quality, or pose potential hazards. The appraiser might evaluate trees to be preserved in terms of the cost to install screening and/or to provide similar aesthetic benefits. Trees to be cut may be evaluated in terms of potential **stumpage value**. A stumpage analysis may require deducting the cost of stump and debris removal. Stumpage value may not exist where there is insufficient volume to cover operating costs, or where there is no evidence that a typical buyer would contemplate selling the timber.

Principle of Conformity

The **principle of conformity** holds that property value is created and sustained when the property characteristics conform to the demands of its market as expressed in economic pressures and shared preferences for certain types of structure, amenities, and services. Land use regulations also influence standards for conformity (Appraisal Institute 2013a).

As an example, a lower-priced property will be worth more in a neighborhood having higher-priced properties than it will in a neighborhood having similar properties. Similarly, a higher-priced property will be worth less in a lower-priced neighborhood than it will in a neighborhood of similar properties. The same principle applies to plant and landscape values. A Japanese red maple in excellent condition will probably add more value to a well-maintained high-value property than to a poorly maintained low-value property.

Unique Aspects of Plant Appraisal

Plant appraisal focuses on trees, shrubs, and other plants as well as associated physical features and maintenance services. Potential assignments are diverse, ranging from one or more trees which may be naturally occurring or planted to project sites that may be income or non-income producing.

The common theme across all plant appraisal assignments is the focus on plants, particularly trees. As assets to be appraised, trees and other woody plants have unique features which differ from other objects of valuation.

Plants are diverse in size, form, and character. There is great variation in plants subject to appraisal, from long-lived trees to shrubs and ground covers to herbaceous plants. This is in stark contrast to commercial and residential land, houses, structures, and equipment, which are relatively easy to categorize. Plant appraisal, like art and gem valuations, requires specialized knowledge. It involves the additional complexity of being part of a larger real estate property whose features may also need to be considered.

Plants provide a wide range of benefits and utility. As described in Chapter 1, trees and other plants provide significant aesthetic, ecological, and environmental benefits, such as fruit production and wildlife habitat, and social and psychological benefits, as well as contributing to real estate market value. Identifying these benefits is much easier than quantifying them. Programs like i-Tree Eco assist the appraiser in understanding the intensity and scale of some (but not all) benefits.

Trees and other plants may detract from a property and its overall utility owing to poor health and structure, conflicting with infrastructure, fruit and litter production, or invasive status.

Trees are often viewed as a community asset, even though they may exist primarily on private property. What may be good for the property may not always be in the best interests of the community, and vice versa. Trees may be subject to laws and regulations, particularly related to species selection and removal.

Trees can be assets to both the community and private property owners. Street trees that are within a public right-of-way and managed by the community often increase the value of the adjacent property. In Portland, Oregon, U.S., for example, street trees in front of a residence increased the sales price by $8,870 (Donovan and Butry 2010). This benefit may extend to properties near the street tree.

The match between plant species and growing site determines the extent of benefits and overall utility produced by the landscape. Under ideal circumstances, a plant selection for a specific site should maximize benefits and minimize liabilities. For example, a species producing fruit and litter may be regarded as a nuisance and liability when installed in a location near sidewalks, pools, and/or parking. Were the same species to be installed in a planting bed or lawn, the fruit would likely not be considered problematic.

A poor match of species to site is likely to result in reduced benefits and utility. When a large-growing tree lacks adequate growing space, conflicts with adjacent buildings, properties, and infrastructure will reduce its overall utility. Such a tree may never reach its anticipated height and spread, either because environmental conditions prevent growth or routine maintenance reduces its size. This situation reflects a type of superadequacy due to large size.

Plants are living organisms that change over time. Trees increase in height, girth, and crown volume. Some species exceed 100 feet (30.48 m) in height and spread. As a general rule, the utility and benefits provided by a tree increase with canopy size. On the other hand, trees that become too large for the available growing space may detract from value or become liabilities, such as in cases involving conflicts with infrastructure or the screening of desirable views. McPherson (2007) describes how an understanding of tree benefits and liabilities can be useful in management decisions.

In nature, trees live from tens to thousands of years. Useful life spans tend to be shorter in planted settings than in natural ones. An understanding of useful life spans will assist the plant appraiser in assessing the potential for a tree to remain an asset into the future. For example, a 20-year-old purpleleaf plum (*Prunus cerasifera)* is probably nearing the end of its useful life span, even if the individual plant is in good condition.

Trees tend to grow rapidly for a period, then slow in height and spread growth as they mature. Long-lived species may be stable for many years. Ultimately, all plants, whether woody or not, will enter a period of decline leading to death. During this period of decline, the utility and benefits provided may decrease while maintenance costs increase.

Because trees and shrubs are relatively small when purchased and planted, they may not immediately provide many (or any) benefits. There is an expectation of increased benefits as these plants mature. For example, a property owner wants an 8-foot (2.5-m) screen provided by a hedge of evergreen shrubs. The plants may only be 2 feet (61 cm) tall at planting with little screening benefit. As the shrubs grow, the benefits increase. If the plants become too tall, the screening benefit may be offset by increased shade, the cost of pruning, or the loss of lower branches.

Crowded growing conditions may reduce the function and utility of an individual plant. If managed early in development, selective removal to reduce the overall number of plants will allow the retained trees and shrubs to develop, and thereby increase their benefit and utility.

Plant appraisal often focuses on the cost to reproduce benefits and utility. Whereas the appraisal of other types of assets generally concentrates on market value, many plant appraisal assignments involve trees and landscapes that are not commonly exchanged in the marketplace. Transactional data are readily available for residential and commercial real estate and for personal property like artwork, antiques, or house pets. Plant appraisal is unique in that it may require an examination of contributions to a larger property, and empirical evidence to support the appraiser's estimate may be challenging to acquire.

Because transactional data are not available for established trees and landscapes, plant appraisal has traditionally focused on estimating costs to reproduce plants or functionally replace their benefits and utility. Because the market for relocating large mature plants is limited, plant appraisers have relied on data from nurseries and landscape contractors. The trees and other plants that are traded in this marketplace are small in size, especially when compared to larger specimens that are being appraised.

CHAPTER 3

The Appraisal Process

CHAPTER OUTLINE

Overview

Appraisal is a systematic process that uses both quantitative analysis and qualitative judgment to develop and communicate an assignment result of either a cost or a value. This chapter provides plant appraisers with an overview of the plant **appraisal process**, a systematic series of steps which assist the appraiser in developing answers to a client's question about value (Appraisal Institute 2013a, 2013b):

1. Define the appraisal problem and the type of cost or value to be estimated.
2. Prepare a scope of work (i.e., the assignment).
3. Collect relevant data.
4. Analyze the data and apply the appropriate approaches, methods, and techniques.
5. Reconcile the analyses to produce the assignment result.
6. Prepare a report.

These steps are listed in sequence, but an appraisal assignment may not proceed in such a linear manner. Defining the appraisal problem is always the first step, yet many details may not be identified until later. In typical appraisal assignments, steps two and three may overlap. Even the definition of the appraisal problem may require a site visit and/or research. For example, the number, size, species, condition, and exact location of appraised plants may not be discerned without field inspection. Under normal circumstances, steps one, two, and three should be undertaken before steps four, five, and six.

Table 3.1 Questions that will be answered during the appraisal process. Each answer provides context for the problem and assignment. As more questions are addressed, the context becomes more specific.

Who
- Who will be preparing the appraisal?
- Who will be collecting the data?
- Who is the owner of the tree/landscape being appraised?
- Who is the client?
- Who are the intended users of the appraisal?
- Who are any other parties involved in this situation?

What
- What are the characteristics of the item being appraised?
- What is the intended use of the appraisal?
- What is the definition of the cost or value to be estimated?
- What approaches, methods, and techniques are relevant to the problem?
- What data are needed to develop the appraisal?
- What are limitations to developing the appraisal?
- What are the relevant dates associated with the appraisal?
  (effective valuation date, inspection date, date analysis was completed, date report was completed)

Where
- Where is the item located? (community, neighborhood, site description)
- Where is the item present, or has it been moved elsewhere?

Why
- Why is the appraisal needed?
- Why were specific approaches, methods, or techniques employed?

The appraisal problem, assignment, and context can be clarified during the initial contact with the client (Table 3.1). A sample client contact form can be found in Appendix 7.

Following initial contact with the client but before entering a formal agreement, the plant appraiser must decide whether to accept the assignment. Are the appraiser's qualifications and experience appropriate for this assignment? Are the scope and terms of the assignment acceptable? Is the appraiser able to bring confidentiality, competence, objectivity, impartiality, due care, independence, and integrity to the assignment? Is the appraiser able to meet the deadlines and timeline? If not, declining the assignment may be the best course of action.

A plant appraiser's responsibility is to provide an independent, objective, and impartial result without discrimination or accommodation of personal interests. The client provides information and context to the assignment but should not direct the appraiser's work. If an appraiser is perceived as having a personal interest, bias, or a conflict of interest, it may be necessary to step away from the assignment or, at a minimum, to disclose the conflict at the onset of the assignment or as soon as discovered. The CTLA strongly discourages advocacy and other practices that are not rooted in economic reality and empirical data.

Step One: Define the Appraisal Problem

The first step in the process is to identify the appraisal problem. This sets the parameters of the assignment and eliminates ambiguity about the nature of the assignment. In this step, the appraiser identifies

a) the client and intended users of the appraisal,

b) the intended use of the appraisal,

c) the type and definition of the assignment result,

d) the effective date of the appraisal,

e) the relevant characteristics of the tree or landscape being appraised, and

f) any assumptions and/or limiting conditions.

A. Identify the Client and Intended Users

The appraiser's client is the person or entity for whom the appraiser is conducting the valuation. The client may be a landowner, attorney, insurance company, potential buyer or seller, lender, government agency, non-governmental agency, or other entity. Identifying the client allows users of the appraisal to know who has hired the appraiser and to whom the appraiser owes a **duty of care**.

An intended user is a person or entity who will use the information in the appraisal (The Appraisal Foundation 2016a; Appraisal Institute 2015). The client is always an intended user of the appraisal. The client may specify other intended users, such as an attorney, municipality, lender, neighbor, or other party.

Many situations involve people other than the client. They should also be identified as early as possible. For example, a tree is located on a property line and is thus jointly owned; an insurance matter will involve one or more insurance companies, policyholders, and claimants. A lawsuit will involve attorneys, a defendant, a plaintiff, and others. Access to the property and tree may require interaction with a renter or real estate agent. Identifying all relevant parties will help to avoid conflicts of interest and delays in performing the scope of services.

B. Identify the Intended Use of the Appraisal

The intended use of an appraisal answers the question: why does the client need the appraisal? The appraisal may have more than one intended use. Probing questions may be needed during the initial inquiry to help both the appraiser and client identify the intended use(s) and establish the scope of work.

For example, a client may tell the appraiser, "A neighbor came onto my property while my family was on vacation and cut down one of my trees." The situation creates the need for an appraisal, but it does not clearly identify the intended use of the appraisal. Examples of intended use in this situation include the following:

- The client wants to sue the neighbor for the loss incurred.
- The city arborist asked that the client obtain an appraisal to establish the fine the neighbor has to pay for violating the municipal tree ordinance.
- The client wants to know if there has been any change in the value of the property.
- The client wants to deduct the loss from his or her income taxes.
- The client's home-owners insurance company (or the neighbor's insurance company) needs an appraisal to establish the value of the client's claim.
- The client wants to know how much it would cost to replace the lost tree with an exact duplicate.
- The client wants the neighbor to replace the tree with landscaping that will provide the same screening.

Understanding the intended use of the appraisal is critical to identifying and defining the appraisal approach (Table 3.2). The credibility of an appraisal is always judged or measured in terms of the intended use of the appraisal:

- Where restoration costs are of primary concern, cost-based estimates may be most relevant.
- Where superadequacy is a factor, a replacement cost estimate may warrant more weight than a reproduction cost estimate, because it focuses on the concept of utility or benefits provided.
- Where market value estimates are sought, the sales comparison approach may be strongest if transaction evidence is prevalent and if it is important to tie the valuation to overall property value.
- Where the client requests both market value and the cost to restore the site to its predamage condition, it would be reasonable to present both estimates.

Table 3.2 Situations wherein a plant appraisal is requested.

Situation	Comments
Loss or damage associated with ▪ Chemical/pesticide injury or encroachment (e.g., drift, spill) ▪ Pruning without permit or permission ▪ Removal without permit or permission ▪ Fire/flood/natural disaster ▪ Construction activity ▪ Vehicle accidents ▪ Tree failure ▪ Trespass ▪ Vandalism ▪ Landscape installation and/or maintenance	Can range from a single tree to many acres of trees. May involve litigation, insurance claims, or fines.
Eminent domain	The taking of private property for public use, most commonly roads, public buildings, utilities, and public facilities.
Land use approvals during development	May encompass bonding or mitigation of damage for trees planned for retention during development. Municipal codes and/or relevant ordinances may suggest specific direction.
Tree inventory	Records of individual tree attributes within a geographic area. They can be for private, municipal, university, commercial, or other properties. Inventories can include estimates of replacement cost, market values, and/or benefits such as ecosystem services.
Insurable value	A cost or value estimate that is related to insurance coverage.
Insurance claims	Prepared in support of insurance claim resolution.
IRS casualty loss	Tax deduction associated with a property's loss in market value caused by a sudden, unusual, and unexpected event.

C. Identify the Appropriate and Relevant Type and Definition of the Assignment Result

Once the intended use of the appraisal is determined, the appropriate and relevant assignment result can be identified. No matter which appraisal approach is employed, the assignment result will always be a cost or a value.

Cost estimates may include one or more of the following:

- Reproduction cost, i.e., the cost to install a duplicate or replica of a tree or landscape item. It may or may not include depreciation for benefits, function, or utility.

Table 3.3 Key to identifying potential appraisal approaches and methods.

Step	Question	Response
1	Is there a controlling authority that requires a statutory value? Does the authority specify the approach, method, and/or technique to be applied?	If yes, apply the defined methodology.
2	Is the subject property part of a natural or managed forest?	If yes, go to Additional Applications. Cost Approach, Sales Comparison Approach, or Income Approach may be appropriate.
3	Does the assignment call for an estimate of cost or an estimate of value?	If cost, go to step 4. If value, go to step 9.
4	Can the item(s) to be appraised be repaired to the extent that the postrepair condition will be satisfactory or similar to the predamage condition?	If yes, go to Cost Approach, Repair Method.
5	Does the assignment call for developing a cost estimate to reproduce a replica of the landscape item?	If yes, go to Cost Approach, Reproduction Method.
6	Would it be practical for the item being appraised to be reproduced by planting a new tree or trees of the same or similar size?	If yes, go to Cost Approach, Reproduction Method, Direct Cost Technique.
7	Does the assignment call for estimating the cost to reproduce a tree that is too large to be planted?	If yes, go to Cost Approach, Extrapolated Cost (Trunk Formula Technique or Cost Compounding Technique).
8	Does the assignment call for developing a cost to replace the item's function or utility in the landscape?	If yes, go to Cost Approach, Functional Replacement. Consider extrapolation (TFT or CCT) or direct cost technique.
9	Does the assignment call for an assessment of change in market value?	If yes, go to Sales Comparison Approach. May consider cost and income approaches as well.
10	Does the assignment call for an assessment of change in market value, based on the cost to reproduce the lost item or its utility?	If yes, go to Cost Approach, then reconcile to market value.
11	Does the assignment call for estimating the present value of future benefits that were lost when the item was destroyed?	If yes, go to Income Approach.
12	Does the item or property being appraised generate income? Crops or commodities include orchards, woodlots, nurseries, vineyards, and Christmas tree farms.	If yes, go to Income Approach. May consider sales comparison and cost approaches as well.
13	Does the assignment support IRS casualty loss valuation?	If yes, go to Additional Applications. Consider cost, sales comparison, and/or income approaches.

Note: This is a simplified series of questions. Not all situations or subtleties are considered. More than one approach, method, and technique may be appropriate for a particular situation.

- Replacement cost, i.e., the cost to replace the landscape item with an item or items having equivalent utility. It may or may not include depreciation for benefits, function, or utility.
- Repair cost, i.e., the cost to repair a damaged landscape item.
- Other costs, i.e., costs associated with the above such as demolition, clean-up and debris disposal, permits, and monitoring.

Value estimates may include one or more of the following:

- Contributory market value, i.e., what the landscape item adds to overall property value.
- Market value of standing or felled timber, crops, nursery stock, Christmas trees, or other commodities.
- Insurable value, which is used to define market value insurance coverage for specific trees or landscape elements.
- Present value of future benefits, such as an i-Tree Eco evaluation.
- Public interest value, i.e., how much the landscape item is worth to the general public.

Plant appraisers may also encounter **statutory value**, i.e., a value specified in a statute or regulation. Statutory value may be calculated according to the number of trees, trunk cross-sectional area, amount of canopy, square foot of cleared area, or some other metric. The method of calculation may be specified. The statutory value calculation may become a fine or penalty for a violation, or a type of mitigation in a project approval. Statutory value is unrelated to an estimate of cost or value.

Definitions of cost and value can be confusing. The definition should be cited from an appropriate source, such as this *Guide*, *Uniform Standards of Professional Appraisal Practice (USPAP)* (The Appraisal Foundation 2016a), *The Dictionary of Real Estate Appraisal* (Appraisal Institute 2015), another text, applicable law or regulation, or an engagement letter.

The plant appraiser selects the approach, method, and technique to develop an estimate of cost or value. A series of questions may assist the appraiser in these decisions (Table 3.3).

Some plant appraisal problems can be addressed in more than one way. An appraiser may determine that more than one approach, method, or technique is necessary to produce a credible cost or value estimate. The assignment may call for simply reporting various results or for reconciling them into a final conclusion.

D. Identify the Effective Date of Valuation

An appraiser should identify the effective date of the valuation in one of three general time frames.

- *Retrospective appraisals* develop an opinion of value that predates the inspection. They are typically retrospective to the date of tree damage or removal. Litigation develops over time, so the appraiser may not become involved until several years after the date of loss, which is likely the effective valuation date.
- *Current appraisals* occur when the effective date of the appraisal is the same as the date of property inspection or report completion, or is current with today's market conditions. They are commonly associated with tree inventories, tree protection or construction bonds, transactions, and damage claims.
- *Prospective appraisals* offer an opinion of value for a future date. They may be appropriate in the case of planned development or as part of cash flow analysis in which projections of future cost or value are made. They may also be used to estimate the value of a nursery, orchard, or timber resource at some future date.

E. Identify What Is Being Appraised

The appraiser needs to identify what is being appraised:

- Is it one or more trees, shrubs, an orchard, turf, timber (standing or cut), hardscape, or a combination of these?

- Is the item real property or personal property? A tree being held in a temporary location, with the intent to plant it in the future, is personal property. A tree growing in a landscape is real property.
- Who owns the property? Is the client the tree owner or the tree owner's representative? What is the ownership interest in the asset(s) being appraised? Are there any limitations to the ownership of the tree or landscape due to easements, leases, timber rights, or a tree preservation ordinance?
- Where is the tree or landscape located? Identify the community, neighborhood, and address. State where on the site the plant or landscape item is placed. Is the tree on the property line or adjacent to a street, streetlight, sidewalk, etc.?
- How is the property currently being used? For example, is it a residence, a vacant lot, a commercial site, an urban street setting, a park, or a woodland? Is a different use planned or proposed? Is the current use the highest and best use?

The characteristics identified in defining the appraisal problem should be sufficient to describe the problem and assignment. Many details may not be identified until later. For example, the number, size, species, condition, and exact location of appraised plants may not be ascertained until actual field inspection and data collection have been completed.

F. Identify Assumptions and Limiting Conditions

All appraisals include **assumptions** and limiting conditions that the appraiser should communicate in the report so that users will understand how to apply the results. Appraisers often undertake assignments with the understanding that some of the information being presented, and upon which the appraiser's conclusions are based, is presumed to be true but has not been verified. Such unverified information is known as an assumption.

Limiting conditions are constraints to the investigation, data analysis, or use of the report. Limiting conditions may be imposed by the situation at the site, clients, a controlling authority, or the appraiser.

Assumptions and limiting conditions are statements that are specific to each assignment and report. Examples include the following:

- Access to the trees was denied, so the appraiser had to rely on distant views of the subject and aerial imagery.
- Property lines were not clearly marked, so the appraiser had to rely on the client's representations regarding to where the tree is placed on the site.
- This report does not confer upon the appraiser an obligation to testify or otherwise participate in subsequent litigation proceedings unless or until arrangements have been made to do so.
- The appraiser obtained cost estimates from sources (e.g., nurseries, landscapers) considered to be reliable and presumes that they are accurate. The appraiser does not assume responsibility for the accuracy of information furnished by other parties.
- This report was prepared using forms developed by the Council of Tree and Landscape Appraisers. However, the content, analyses, and opinions set forth in this report are the product of the appraiser.

Extraordinary assumptions are assumptions "directly related to a specific assignment, as of the effective date of the assignment results, which, if found to be false, could alter the appraiser's opinions or conclusions" (Appraisal Institute 2015: p. 3). Labeling them as such signals to the reader that the assumption may not be true, whereas in the case of the more general assumptions noted previously there is no particular reason to caution the reader. Extraordinary assumptions should be clearly disclosed in the appraisal report.

Examples of extraordinary assumptions include statements such as the following:

- There was no opportunity for the appraiser to observe the health of the trees before they were damaged. The appraiser assumes that they were reasonably healthy at the time of loss.
- The trees that were cut were removed from the site before the appraiser's inspection. The appraiser assumes that the correlation between stump diameter and dbh observed among similar trees remaining on the site is indicative of the correlation for the trees that were removed.

- The appraiser is not aware of any local history of purchasing and installing large tree specimens like the subject tree. The client nevertheless has directed the appraiser to estimate the cost to install a similar specimen, under the assumption that it is economically feasible.

Hypothetical conditions are assumptions made contrary to known fact, but which are regarded as true for the purpose of discussion, analysis, or formulation of opinions (The Appraisal Institute 2013a). Hypothetical conditions should only be used for legal purposes, for purposes of reasonable analysis, or for purposes of comparison, and if they lead to a credible analysis. Hypothetical conditions should be clearly and conspicuously disclosed in the appraisal report.

Statements of hypothetical conditions include the following:

- The subject plants were in poor to average condition at the time of the damage. However, the client has instructed the appraiser to estimate their reproduction cost absent any deductions for depreciation. The resulting estimate therefore exceeds the estimate that would be produced if depreciation were considered.
- The subject tree is an invasive species and, in the appraiser's opinion, was undesirable as a landscape component. There is no evidence that property owners planted such specimens. However, the client has asked the appraiser to appraise the reproduction cost of the subject tree without deducting for its invasive status.

Step Two: Define the Scope of Work

Once the appraiser has identified the appraisal problem, a scope of work can be determined. It should contain a statement of the appraisal problem, the type of report to be provided, a time line, and a fee. The client and appraiser should agree on the scope in writing prior to proceeding with the project. The scope of work may accompany a contract or it may be the contract. The following are suggested steps to developing a scope of work:

1. Define the appraisal problem.
2. Schedule a date and time to inspect the property and collect field data.
3. Identify the need to research background information such as nursery tree costs, property values, or income data.
4. Identify the approach, method, and techniques to apply.
5. Disclose appropriate assumptions and limiting conditions.
6. Define the type of, and due date for, the appraisal report.
7. Estimate the price for the work and present a payment schedule.

The scope of work should be sufficient to provide credible results supported by relevant evidence and logic. It should not, however, be excessive or go beyond the bounds of the assignment. For instance, if the assignment result focused on a reproduction cost, obtaining data about the highest and best use of a property or overall property value would not be necessary.

In addition to defining the appraisal problem and the scope of work, the appraiser should identify other assignment information or parameters:

- relevant people or parties
- relevant dates (date of damage, date of hiring, inspection or report deadline, etc.)
- additional services (conferences, meetings, deposition, trial testimony, etc.)

Step Three: Collect Relevant Data

In this step, the appraiser starts to perform the scope of work. Deciding what data are relevant is often a task that spans several steps in the appraisal process. Once the scope of work and assignment have been defined, the

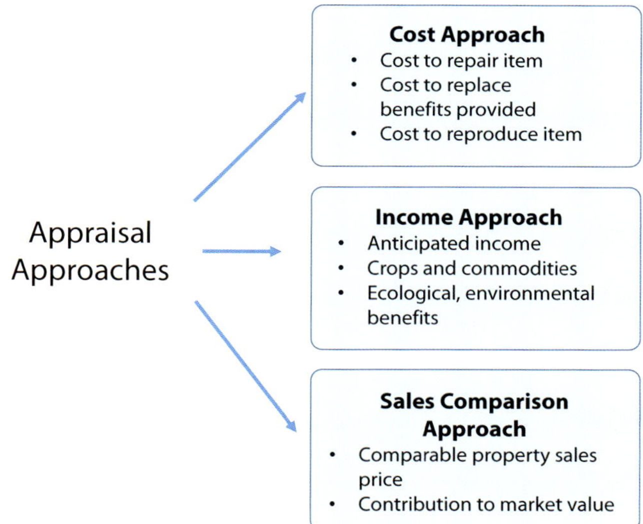

Figure 3.1 Appraisal approaches: cost, income, and sales comparison.

type of data to be collected should be clear. It may be necessary to inspect the property or complete initial research to identify the data that are actually available.

Appraisals intended to be used for resolving litigious situations will likely require discussions with attorneys, clients, and other parties. Such discussions may lead the appraiser to research local ordinances, state statutes, past appellate case rulings, or contractual language.

Step Four: Analyze the Data by Applying Relevant Valuation Approaches, Methods, and Techniques

In this step, the appraiser continues to perform the scope of work by analyzing the collected data. There are three appraisal approaches: cost, income, and sales comparison (Figure 3.1). Each one involves one or more methods, and each method may involve one or more techniques.

The cost approach (Chapter 5) analyzes the costs of goods or services to estimate (1) the cost to repair the item, (2) the cost to replace the benefits provided by the item, or (3) the cost to reproduce the item. The cost approach is based on the principle of substitution. That is, a buyer would not pay more for an asset than the cost to acquire an asset with similar utility.

The income approach (Chapter 6) develops a value for income-producing assets by estimating anticipated income. Plant appraisers have traditionally used this method to estimate different types of value for green-industry businesses such as nurseries, Christmas tree farms, orchards, timber, etc. In addition, the present value of current and anticipated ecological and environmental benefits can be quantified with programs such as i-Tree Eco. The income approach is based on the principle of anticipation, or the expectation of future benefits (income) on an annual or recurring basis.

The sales comparison approach (Chapter 7) examines comparable property sale prices. For plant appraisal, the appraiser is most commonly concerned with how much the plants in question contribute to the market value of the overall property. The sales comparison approach is based on the principle of substitution, the theory that knowledgeable buyers acting in their own best interest would not pay more than the price necessary to obtain a comparable property.

Before selecting a specific approach and method, the plant appraiser should determine if there are local, state, federal, or other controlling authorities that require a specific approach and/or method. For example, in the state of Florida, methods for determining the mitigation value of roadside vegetation are described in State Administrative Code (Rule 14-040.030).

The result of data analysis is called the assignment result, which is an estimate of cost or value. When the assignment result is a cost, it should not be called an appraised value.

Step Five: Reconciliation

Reconciliation, when necessary, is the final step in developing the assignment result. USPAP requires reconciliation with every appraisal, even where only one of the three approaches is applied (The Appraisal Foundation 2016).

During reconciliation, the appraiser weighs the strengths and weaknesses of each approach, method, and technique used based on the quality of data, the level of subjectivity of the analysis, and the relevance of the approach to the appraisal problem. It is not a mathematical process. An appraiser relies on professional experience, expertise, and judgment more in reconciliation than in any other part of the appraisal process.

Reconciliation involves two steps. First, the processes that led to the different value indicators are reviewed and checked for errors. Second, the reconciliation judgment is made.

Step Six: Prepare the Report

The final step in the appraisal process is communicating the assignment result and supporting information obtained throughout the appraisal process to the client and other intended users (see Chapter 8). This can be done verbally or, more commonly, in a written report.

USPAP recognizes appraisal reports and restricted appraisal reports. An appraisal report provides either a full narrative or a summary, going into sufficient detail to lead the reader through the appraisal process. A restricted appraisal report is very short and merely states key elements of the assignment and the assignment results, such as with an abbreviated tree inventory.

An oral report is any verbal communication that transmits the assignment result from the appraiser to the client or another intended user. USPAP requires appraisers to support oral reports by creating and maintaining file notes that substantiate their conclusions.

The assignment result will be a cost or a value. Because the two are distinct from one another, the appraiser must be clear about the distinction throughout the entire appraisal process. If there is any chance that the reader might confuse a cost estimate with a value estimate, the appraiser should take care to communicate clearly which of the two the assignment result represents.

Summary of the Appraisal Process

Appraisal assignments may vary but the basic framework for the appraisal process should not. A well-defined appraisal leads to a clear scope of work and informs the selection of approaches, methods, and techniques. The appropriate data is collected and analyzed. A final result is obtained, documented, and reported.

The Appraisal Process—Example 1

A city arborist calls and asks the appraiser to assess the health and structure of a large, mature tree that was recently "trimmed." In addition, the appraiser is asked to determine if the value of the tree changed as a result of the trimming. The appraiser agrees to meet the city arborist at the site. The arborist says that a proposal will not be needed.

When visiting the site, the city arborist provides further details. The tree is located behind a local tavern. The base of the trunk is 1 foot (30.48 cm) from the property line. The adjacent property is the parking lot of a bank. All of the branches extending over the parking area were headed back to the property line (indicated by the red circle in the photo).

The appraiser collects relevant information about the tree: health, structure, form, diameter, crown size, extent of pruning, and surrounding landscape. The tree meets the City's criteria for heritage status. The tavern owner relates that the tree is an integral part of the rear-yard landscape but requires daily clean-up of falling leaves, twigs, fruit, and other debris. When viewed from the tavern side, the effects of the trimming on the tree's appearance are minimal.

The appraiser informs the city arborist that the tree does not seem to be fatally damaged. The appraiser describes the need to prune the tree to correct some of defects created by the trimming. The city arborist says that the City will fine the owner of the bank property, as they engaged the landscaping firm that performed the trimming, and asks the appraiser to estimate any loss in value.

The appraiser determines that using the cost approach and the trunk formula technique will produce the most reasonable assignment result. The estimated loss in value associated with the trimming is the reproduction cost prior to the trimming minus the reproduction cost after the trimming. In addition, the cost of the restoration pruning is also relevant. The appraiser finds an aerial photograph of the tree dated six months prior to the trimming on the internet and assesses the condition of the tree to be good prior to the trimming and poor after the trimming.

The appraiser prepares a letter report describing the assignment, the information collected in the field, the method and assumptions used to estimate the loss in value, the assignment result, and recommendations for pruning.

Review of the Process

1. Define the appraisal problem and the assignment result.	A heritage tree had been severely trimmed. The city arborist (the client) planned to fine the offending party for any loss in value and cost of restoration pruning. To address this problem, the appraiser first needed to determine if the tree should be removed or if it could be retained. The assignment result included an estimate of value (based on the estimated reproduction cost before and after the trimming) and an estimate of the cost to prune the tree to restore the crown.
2. Define the scope of work.	No formal proposal was required. But the scope of work included meeting at the site, collecting relevant data at the site, collecting relevant data about the tree prior to the trimming, identifying reasonable cost information, estimating the loss in value, defining the pruning required, preparing pruning guidelines, and then preparing a report.
3. Collect relevant data.	Data was collected and evaluated in the field and office.
4. Analyze the data and apply the relevant valuation approach(es).	The focus on the before and after conditions of the tree and the assessment that it could be retained led to using the cost approach and the trunk formula technique, one of the extrapolated replacement cost methods.
5. Reconcile the analyses.	No reconciliation was necessary.
6. Prepare a report.	The final step.

Summary

This type of appraisal assignment is common for arborists, whether they are involved in the municipal, commercial, utility, or consulting aspects of the industry. A tree protected by a city statute was inappropriately pruned. The city statute calls for the fine to be determined using the methods found in the *Guide for Plant Appraisal*. Because the tree was still standing and accessible, there were no issues collecting relevant data. Historical imagery was easily accessed. The use of the trunk formula technique to establish the difference in depreciated reproduction cost due to the trimming provided the City with the basis to determine the amount of the fine.

The Appraisal Process—Example 2

A consultant is contacted by an attorney regarding a case of chemical trespass and the associated death of several trees, shrubs, and understory plants. During the initial phone call, the general facts and nature of the case are introduced. The parties and attorneys are revealed so that the consultant can perform a check for conflicts.

Following the initial phone conversation, the consultant provides a simple engagement letter detailing the fee arrangement as well as a curriculum vitae. An initial meeting at the tree owners' property takes place the following week.

Background. The tree owners had lived at the property for over 20 years. They had an agreement with the uphill neighbor to maintain a hedge of Leyland cypress (x *Cupressocyparis leylandii*) trees at a height of 25 feet (7.6 m). The hedge was located in the rear yard near the property line. The agreement had been in place for many years and the cost of pruning the trees had been shared by both parties.

The clients' gardener observed that the cypress trees were dying. The lower trunks and soil were coated with a black, oily substance that gave off a very strong odor. Soil test results indicated very low pH with positive indications of a strong acid. A plant diagnostic laboratory concluded that concentrated acid had been applied to the soil and lower trunk of the tree. The county agriculture department agreed with the conclusion that concentrated acid had been applied.

The tree owners filed suit against their neighbors. The attorney wanted to retain the consultant's services as an expert with the intention to call that person as an expert witness at the trial.

Step One: Defining the appraisal problem and the assignment result. After the initial contact and site visit, the consultant defines the appraisal problem. The matter is in litigation and involves chemical trespass. There is evidence to support the assertion that the trees and other landscape plants have been killed by a strong acid. The property owners want to restore their landscape, i.e., to reproduce the landscape as it was prior to the application of the chemical. The appraisal will be used to support a claim for damages.

Step Two: Establishing the scope of work. Based on conversations with the attorney and tree owners, the following scope of work is established:

1. Review documents, photographs, and laboratory test results provided by the attorney.
2. Inspect the damaged landscape to document the species, condition, number, and size of affected plants; establish the size of the affected area; take photographs; and make any other relevant observations and measurements.
3. Develop a conclusion as to the cause of tree death and assess the likelihood that additional plants will decline and die in the future.
4. Develop an estimate of the cost to restore the landscape to its prepoisoning state.
5. Prepare a report if requested.

Verifying the location of the property line is not included in the scope of work. Until otherwise determined, the consultant assumes that the fence separating the two properties represents the property line. The location of the property line has not been in dispute in all the years of maintaining the trees nor is it raised by the lawsuit.

The scope of work is established through conversation and agreement with the attorney. No formal proposal or contract is prepared.

Step Three: Collect relevant data. Data collection proceeds in three steps:

1. Review reports, documents, statements, and photographs regarding the timing, cause, and pattern of the damage. This includes a review of soil-test reports.
2. Inventory the affected area. Identify the species, size, condition, and location of the dead and damaged plants. Determine which, if any, plants can be repaired. Note site layout and topography. The clients'

property slopes uphill from the street. The rear yard is only accessible from the street by a set of stairs. The distance between the house and its side yard property lines is less than 30 inches (75 cm).

3. Obtain cost information. The consultant contacts a landscape contractor familiar with the area and type of landscape. They meet on-site to discuss the steps required to restore the property. Among other observations, the contractor determines that all of the work needs to be performed by hand, since there is no access for equipment, and a crane cannot be used to move plants and material from the street to the worksite. The consultant also checks the availability and price of the needed plants with two local wholesale growers.

Three site visits are made: (1) the initial meeting, (2) a data collection visit, and (3) a meeting with the landscape contractor.

Step Four: Analyze the data and apply appropriate approaches, methods, and techniques. The dead cypresses were 25 feet (7.62 m) tall and formed a continuous hedge along the property line. A nursery-grown 24-inch (0.6-m) box cypress is the largest tree that can be moved by hand to the rear yard. Such a tree would be 10 feet (3.05 m) tall and 3.5 feet (1.07 m) wide. At that size, trees could be placed box-to-box and offer some immediate screening but would be smaller than the poisoned trees. Contaminated soil also has to be removed and replaced.

The appraisal consists of two parts. First, all the costs for labor and materials are totaled. Second, cost compounding is applied to the cypress trees. The consultant estimates it would take ten years for newly planted trees to establish a continuous screen at 25 feet.

Step Five: Reconcile the analyses to produce the assignment result. Reconciliation consists of reviewing the data, methods, and calculations. The combination of direct cost and cost compounding is discussed with the attorney.

Step Six: Prepare a report. No written report is prepared. An oral report is provided to the attorney. The case settles at mediation.

This project was typical of the consultant's litigation experience. Little or no paperwork was exchanged. No report was prepared. Were this not a matter of litigation, a formal proposal, scope of work, and contract would have been prepared, and a report would have been submitted.

For a chemical trespass case, this project was very straightforward. The assignment did not address issues of intentional acts or any wrongdoing. There were no issues over the location of the property line in relation to the fence.

The appraisal process was also straightforward. Limited access made the restoration more time consuming, and it was a significant limiting condition.

Note that only four of the six steps were completed. Had a report been required, it would have described how and why the cost approach and reproduction method were chosen and why they were applicable to this setting.

Data Collection

CHAPTER OUTLINE

Overview

Appraisers form opinions of value that are based on data. Data are facts that provide the basis for reasoning, discussion, or calculations. The exact data needed for an appraisal will vary with the assignment, approach, method, and technique applied.

Appraisers collect both quantitative and qualitative data. **Quantitative data** are measurements, counts, rankings, or ratings (e.g., height, diameter, and number of trees). **Qualitative data** are descriptive, relating to qualities (e.g., plant health, site conditions, and management history). The subjectivity of qualitative data does not render it any less important than quantitative data. For example, veneer-grade trees and Florida fancy-grade nursery stock are commodities where grade (quality) affects selling price.

Categories of data that are important to many plant appraisals include the following:

- site information,
- observations of plant or landscape features, and
- off-site research.

Recording and Managing Data

Dunster (2014) describes evidence as the facts that support an approach, argument, reasoning, or conclusion. Data, whether field measurements, observations, or research, establish those facts. A plant appraiser's understanding of the appraisal problem is key to determining what information needs to be obtained and how it can be obtained. In addition, Dunster argues that the manner in which evidence is collected and organized can facilitate good reasoning and judgment.

Dunster also emphasizes that comprehensive knowledge and experience are the keys to collecting and analyzing evidence. In his view, there are two fundamental principles to good data collection. First, the plant appraiser should know what to look for. Second, the plant appraiser should know what they are looking at.

Plant appraisers find a field form to be a useful tool in collecting relevant data (see Appendix 7 for a sample field-data collection form). Such forms can be paper or electronic. In either case, field forms become the starting point for analysis and decision-making. Such forms should be retained with the appraiser's project file. Photographs are another source of information and should also be retained.'

ACCURACY vs. PRECISION

"In simplest terms, given a set of data points from a series of measurements, the set can be said to be **precise** if the values are close to the *average value* of the quantity being measured, while the set can be said to be **accurate** if the values are close to the *true value* of the quantity being measured. The two concepts are independent of each other, so a particular set of data can be said to be either accurate, or precise, or both, or neither" (https://en.wikipedia.org/wiki/Accuracy_and_precision. Accessed August 28, 2017).

Imprecision may be thought of as resulting from random errors while inaccuracy results from systematic errors. For example, under normal circumstances, trunk diameter is measured at 54 inches (137.16 cm) above grade. Accuracy is assessed by how close to 54 inches the average measurement is taken. Precision is the variation from the actual height of measurement.

Site Information

Most appraisals begin by collecting background information on the site where the plants or landscape items are located. A site is the area where the appraisal is being performed and can be, for example, a residential lot, a right-of-way, or a larger parcel of land. Data regarding the site are collected during the initial conversation with the client and throughout the field visit.

Site information may include the following:

- address, legal description of land, or surveys;
- current land use (residential, wooded, commercial, industrial, park, right-of-way, urban, rural), zoning and property tax descriptions, or assessment records;
- structures (buildings, hardscape, etc.);
- maps, aerial photographs, architectural drawings, or landscape plans;
- topography such as slope and aspect;
- soil type, volume, and drainage;
- climate and environmental data, such as hardiness zones;

ROUNDING TO SIGNIFICANT FIGURES

Data collected in the field may not be exact. Plant appraisers commonly round measurements to something that is shorter, simpler, and (possibly) more explicit. Pi (π) is often rounded to 3.14. Trunk diameters are often rounded to the nearest tenth of an inch (e.g., 14.7 inches) or the nearest inch (e.g., 15 inches). Measurements of tree height are rounded to the nearest foot, or perhaps five feet for tall trees. Rounding is also a normal feature in developing estimates of cost or value.

The appropriate number of significant figures is the minimum number of digits needed to produce results without producing bias or unduly compromising **accuracy**. For whole numbers (not fractions), the number of significant digits is indicated by the maximum number of digits (not including trailing zeros) as in the following table.

Examples of rounding to 3 significant digits

Original number	Rounded to 3 significant digits
1,345,222	1,350,000
987,500	988,000
24,805	24,800
749,998	750,000
1,588.65	1,590
6.4	6.40

Rounding to the nearest one-hundredth is as follows:

Examples of rounding to 2 significant decimal digits

Original number	Rounded to 2 significant decimal digits
100.088	100.09
13.766	13.77
5.2957	5.30
0.506	0.51
0.008	0.01
0.0035	0.00

Any estimate of cost or value should be rounded to reflect the overall **precision** or reliability of the appraisal process as a whole. The greater the number of significant figures presented, the greater the implied precision. The following illustrates the implication of rounding a calculated figure of $75,649 to different significant digits. Using two significant digits implies a precision of ±1%.

For total cost and value estimates, it is rarely necessary to round to more than two or three significant digits.

Implication of rounding $75,649 to different significant digits to precision

Number of significant digits	Rounding result	Implied precision
1	$80,000	1/8, or ±12%
2	$76,000	1/76, or ±1%
3	$75,600	1/756, or ± 0.1%

Rounding to three digits produces the following figures: $107.35 rounds to $107; $8,347 rounds to $8,350; $52,844 rounds to $52,800; and $356,604 rounds to $357,000. For numbers over about $500, rounding to two digits infers precision of 1% or less.

Depreciation in the cost approach diminishes the precision of the overall process. It is meaningless to report a cost conclusion to four significant digits when the reliability of the estimate is more akin to two significant digits.

Rounding should not produce bias. Bias can occur when the appraiser rounds at multiple intermediate points in the analysis because of the potential cumulative effect of iterative rounding. As a general rule, the mathematic impact of rounding should determine the appropriate number of significant digits. An exponent has a much greater impact on a number than a non-exponent, so it should be expressed in very precise terms.

For intermediate calculations, it is reasonable to round numbers like pi to two significant decimals, or 3.14. It is perfectly acceptable to use 3.14159, but generally unnecessary. For cost forwarding, an interest rate reported to one decimal place will normally suffice, primarily because the very process of selecting an appropriate interest rate is subjective. Despite the fact that estimates can be highly sensitive to exponents (particularly over long time horizons), highly precise derivation of interest rates is nearly impossible.

Where multiple valuation approaches are used, it is appropriate to round the conclusion for each approach to the same precision level as will be used for the final reconciled estimate. Doing so reflects the common experience of buyers and sellers, who negotiate in terms of rounded figures.

In general, the plant appraiser should avoid reporting conclusions with superfluous precision.

- historical documents;
- photographs; and
- legal rulings, statutes, and ordinances.

Structures

Plant appraisers may need to understand how landscape items affect the architectural appearance and engineering of buildings on a site, and how buildings affect plant growth, vigor, longevity, etc. Observations regarding proximity and interface between buildings and plants or hardscape items are important for understanding whether plants are used appropriately. The interaction between plants and buildings or structures can be positive, negative, or neutral.

Relevant building characteristics may include type (e.g., house, multi-story, warehouse, skyscraper), age and type of construction (for ascertaining potential root damage, etc.), and footprint relative to property (e.g., limited space for tree growth).

Hardscape

Plant appraisers may need to collect quantitative and qualitative data on hardscape features. An appraiser may need to know if there is a relationship or interdependence between a hardscape element and a plant. Relevant hardscape items may include the following:

- paving systems (walks, patios, drives, etc.);
- fences, walls, and gates;
- retaining walls;
- roads and bridges;
- overhead structures (powerlines, structures, trellises, arbors, pergolas);
- irrigation systems;
- lighting systems;
- landscape rocks and boulders;
- sculptures and garden art;
- water features (fountains, ponds, streams, plumbing); and
- drainage systems (pipes, inlets, catch basins).

In analyzing these features, special knowledge may be required to estimate a cost or value. An appraiser may need to hire other experts, such as a landscape architect, landscape contractor, structural engineer, electrician, plumber, or art appraiser, for the analysis.

Plant Information

Certain data are required for most plant appraisals, including the identification, size, placement, and condition of the plant.

Plant Identification

An appraiser should accurately identify the plants being appraised to the extent required by the appraisal problem. Appendix 1 describes the system used in naming plants.

Plant Size

The dimensions measured depend on the type of plant being appraised and the approach being applied. Turf and ground covers are measured by area (square foot or square meter). Annual and perennial herbaceous plants can be quantified by bed area, plant height, width, and/or plant count. Shrubs are quantified by number, height, spread, and/or volume. Trees are measured by trunk diameter, height, and/or **crown** size. Timber cruises develop an estimate of harvestable wood volume in the form of log diameter and length, which leads to an estimate of board feet of lumber.

Nursery stock measurements are defined by industry standards or regional practices (ANSI Z60.1-2014). Conifers and multi-stem deciduous trees are commonly sold by height. Herbaceous plants, shrubs, and small trees are often sold by container size. In the western U.S., nursery stock is sold by container size. In the eastern U.S., larger nursery stock is generally sold based on either trunk caliper or plant height.

Measurements of plant size are taken on-site or calculated following the site visit. Data from the field, such as planting bed dimensions, trunk diameter, and crown height and spread, are used to calculate bed area, trunk cross-sectional area, and crown volume.

Height

Plant height measurements are often included in the plant description, and they are used when appraising plants that provide screening, that are being grown for timber, or that are sold by height (e.g., shrubs, conifers, palms, Christmas trees). Height is measured from the soil line to the top of the plant. The appraiser can collect height data using a **clinometer**, **hypsometer**, laser range finder, measuring tape, or telescoping rod.

Palms are measured by trunk height or tree height (Figure 4.1), depending on region and species. Plant appraisers should consult local buyer guides or suppliers of palms regarding the locally accepted procedure.

Figure 4.1 Trunk height is measured from the ground line, which should be at or near the top of the root zone to the base of the heart leaf (ANSI Z60-2014).

Trunk Diameter

In the U.S., trees growing in nurseries with trunk diameters up to and including 4 inches (10 cm) are measured at 6 inches (15 cm) above the ground (ANSI Z60.1 2014) (Figure 4.2). When the diameter at 6 inches (15 cm) above the ground is greater than 4 inches (10 cm), the height of measurement is then increased to 12 inches (30 cm).

Trees outside of nurseries are measured at 4.5 feet (1.37 m) above the ground, a point referred to as **diameter at breast height (dbh)** or diameter at standard height (dsh) (Bond 2014) (Figures 4.2 and 4.3a-l). Trees are measured at dbh to reduce the influence of trunk flare and for convenience of measurement. Outside the U.S., trees are often measured from 51 to 59 inches (1.3 to 1.5 m) above existing grade (Swiecki and Bernhardt 2001).

Figure 4.2 The location of diameter measurement varies by location. (a) The caliper of an American elm growing in a nursery is measured 6 inches above the ground when the stem is ≤ 4 inches. When the diameter at 6 inches above the ground is greater than 4 inches, the height of the measurement is then increased to 12 inches above the ground. (b and c) The diameter of the same tree growing in the landscape would be measured at 4.5 feet above the ground (dbh).

Field Adjustments to Measuring Diameter

Because the form of the lower trunk varies widely, it is not always possible to measure at 4.5 feet (1.37 m). Trees with excessive basal flare, leaning trees, trees on a slope, trees with multiple trunks (stems), low-branching trees, and other patterns of tree growth make it necessary to modify where diameters are measured (Figures 4.3a-l). The goal of adjusting the measurement location is to best represent the size of the tree. If the measurement location differs from dbh, the appraiser should record the height at which diameter was measured.

Vertical trunk, level ground. The trunk measurement of a tree growing vertically on level ground should be made at 4.5 feet (Figure 4.3a).

Leaning trunk, level ground. The trunk measurement of a leaning tree on level ground should be made 4.5 feet from the ground on the compression or underside of the trunk. Measurement of the trunk diameter should be perpendicular to the trunk (smallest diameter across the trunk) (Figure 4.3b).

Leaning trunk, slope. The trunk measurement of a leaning tree on a slope should be made 4.5 feet from the ground on the high side of the slope. Measurement of the trunk diameter should be perpendicular to the trunk (smallest diameter across the trunk) (Figure 4.3c).

Vertical trunk, slope. The trunk measurement of a tree growing vertically on a slope should be made 4.5 feet from the ground on the high side of the trunk (Figure 4.3d).

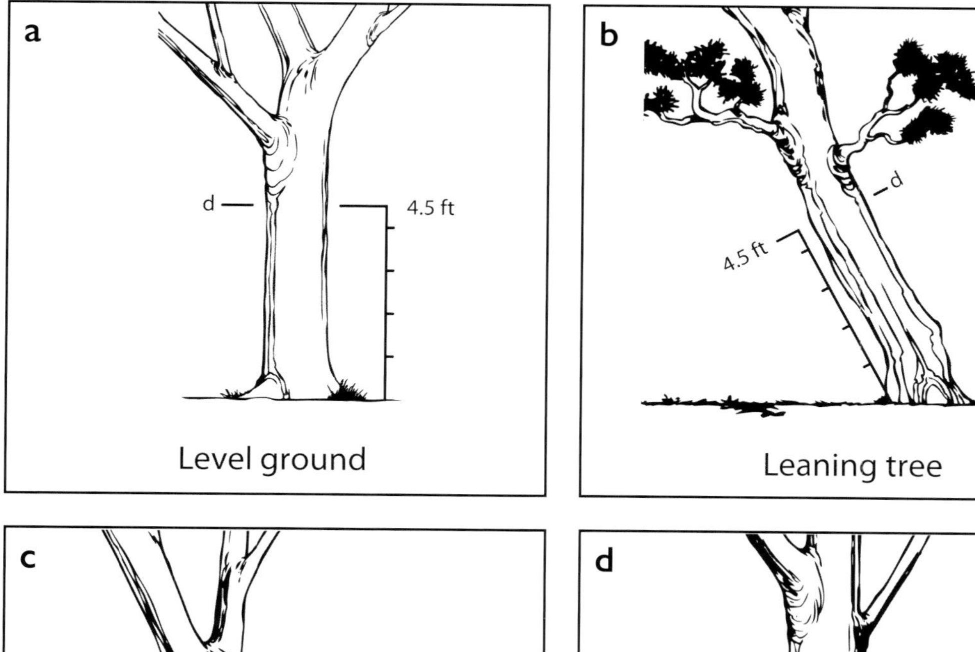

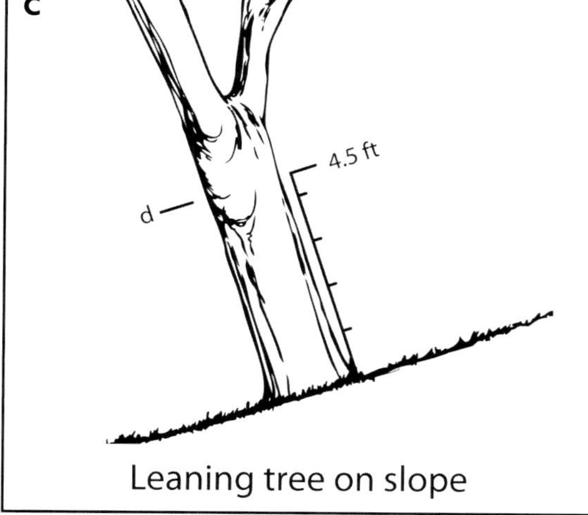

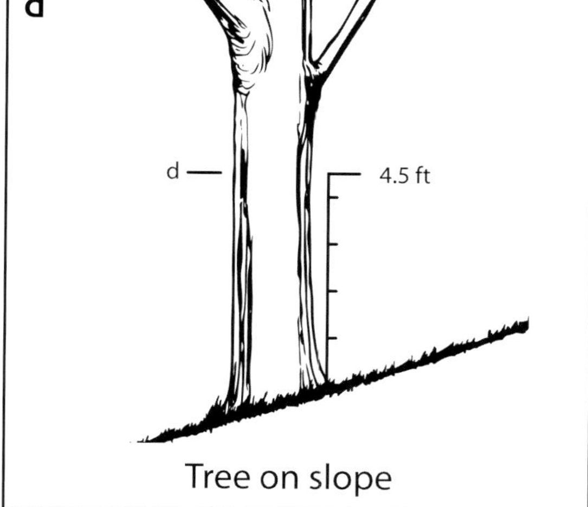

Figure 4.3 Field adjustments to measuring diameter.

Low branching. Low branches may make it difficult to measure at 4.5 feet above the ground. In such cases, measure trunk diameter immediately above the attached branch (Figure 4.3e).

Deformity, swelling, gall, wound. Trunk form may deviate from normal due to wounds, galls, swelling, and other deformities. Measure trunk diameter immediately above the irregularity at the point it ceases to affect the stem (Figure 4.3f).

Trunk with more than one stem originating at or near ground level. If all the stems arise from within 3 feet (1 m) of the ground, and each stem contributes equally to the canopy, then determine the sum of the cross sectional areas of each stem measured at 4.5 feet (1.37 m) above grade (see figure 4.3g). Different stem configurations may require measuring at other heights or locations to more accurately represent the size of a stem (see figures 4.3e-k).

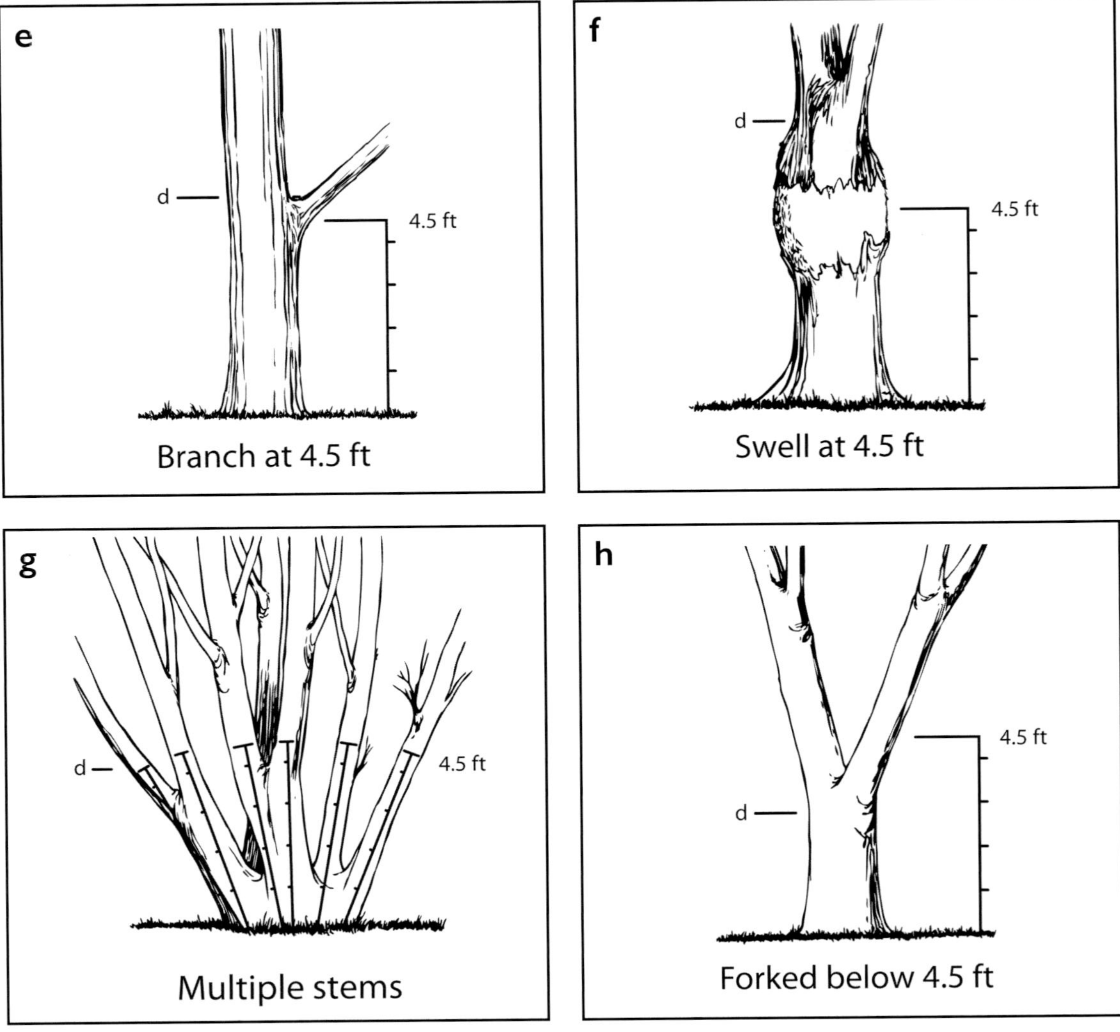

Figure 4.3 Field adjustments to measuring diameter.

Trunk with more than one stem originating from 3 to 5 feet. Where the trunk divides between 3 to 5 feet (1 to 1.5 m), measure below the point of attachment and below any associated swelling (Figure 4.3h). Where trees are being measured for timber and the fork arises below 4.5 feet, measure 3.5 feet (1.5 m) above the bottom of the fork (Figure 4.3i). If the fork arises above 4.5 feet, measure at 4.5 feet (Figure 4.3j).

Excessive trunk flare or basal swelling. Excessive trunk flare presents a problem similar to that of low-branching trees. Measure 1.5 feet (0.46 m) above the neck of basal swelling to provide a reasonable estimate of trunk diameter (Figure 4.3k). Alternatively, measure nearby trees with similar crown size and form and apply that measurement.

Trunk on the ground. Where a tree has been uprooted, windthrown, or is otherwise on the ground, measure 4.5 feet (1.37 m) from the base (Figure 4.3l).

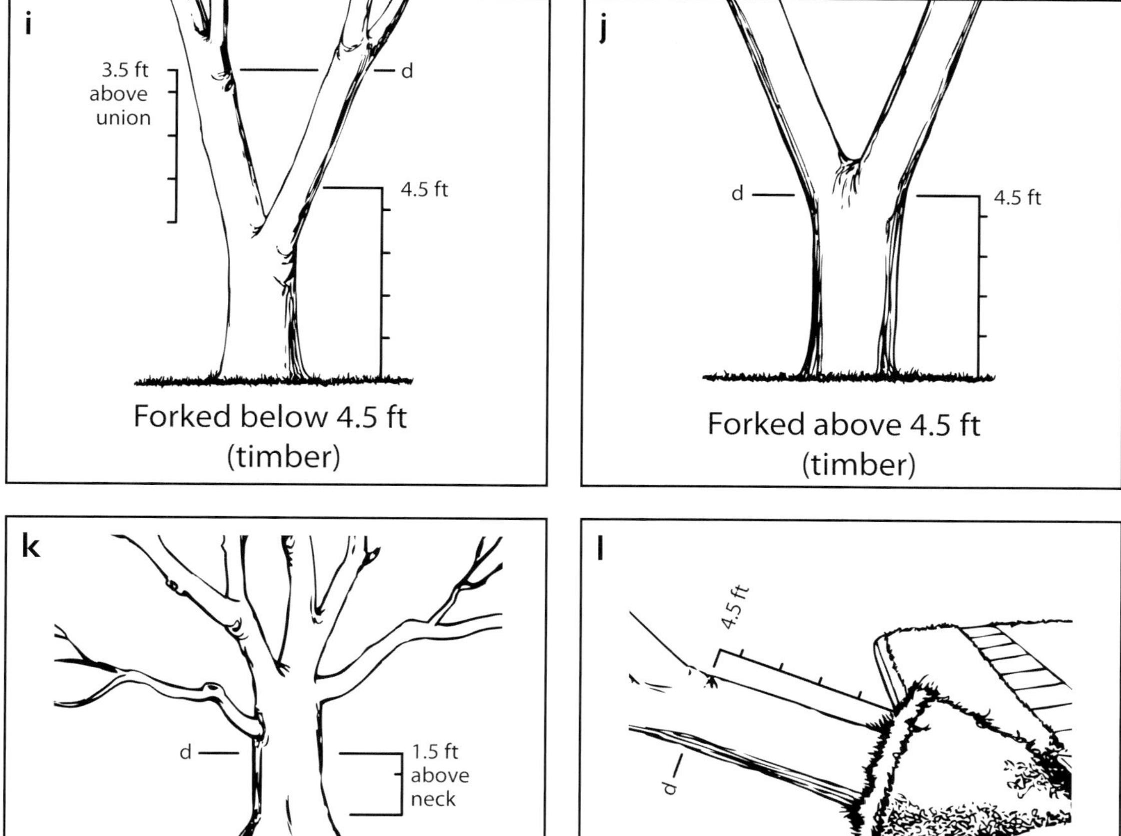

Figure 4.3 Field adjustments to measuring diameter.

Figure 4.4 Tools for measuring trunk diameter: (a) diameter tape with linear scale on one side and diameter on the other, (b) nursery calipers, (c) field calipers.

Substitute Measurements

Sometimes it is not practical to measure trunk diameter, such as when access is not allowed, the tree has been removed, or there is a swelling on the trunk. In these cases, the appraiser has the option to measure nearby trees of the same species and similar size (diameter, height, and/or crown spread), then use that measurement, or the average measurement from several trees, as a substitute. See Appendix 3 for a more complete discussion.

Tools for Measuring Diameter

There are two basic tools for estimating the diameter of a tree trunk: tapes and calipers (Figure 4.4). Diameter tapes are wrapped around the trunk and directly read using the scale marked on the tape. One side of the tape shows lineal distance or circumference, and the other side shows the conversion to diameter (circumference divided by 3.14, or pi [π]). The main advantage of a diameter tape is its ease of use. The main disadvantage is that it overestimates diameter, except where the trunk is a near perfect circle.

Calipers reduce the problem of overestimating diameter by taking two diameter measurements at right angles then computing the square root of their product, known as the quadratic mean diameter (QMD) (Appendix 2). The advantage of using QMD is that it produces a more accurate estimate of cross-sectional area, particularly when tree trunks are elliptical or otherwise asymmetric. Alternatively, one can average the two measurements, though this is less accurate than computing the QMD. The QMD can also be used to calculate the area of the average tree in the stand.

The disadvantages of calipers are that two measurements must be taken and additional calculations are required. Where timber cruising is conducted, recommended practice is to measure the diameter facing the sample plot center. Some measurements will underestimate QMD and others will overestimate QMD, but this random variation will not produce bias.

Figure 4.5 Calculating cross-sectional area of a multi-stem tree.

ADJUSTED TREE AREA (ATA)

The eighth edition of the *Guide* (1992) introduced adjusted trunk area (ATA) as an adjustment for trees over 30 inches (76.2 cm) in diameter. ATA was developed by the CTLA "on the basis of the perceived increase in tree size, expected longevity, anticipated maintenance, and structural safety" (p. 16). The CTLA stated that above 30 inches in diameter, "tree values based on trunk area became unrealistically high" (p. 16). ATA had the effect of modifying the geometric increase in trunk area. For example, the actual trunk area of a 45-inch-diameter (114.3-cm) tree is 1,590 square inches. The adjusted trunk area is 1,353 square inches (see Table 4.4, Council of Tree and Landscape Appraisers 2000).

The ATA formula applied in past editions results in adjusted trunk area decreasing for diameters exceeding 103 inches (261.6 cm). ATA recognizes that the rate at which additional benefits accrue begins to decline with advanced age and increasing diameter. Trees, like most any other organism, age, decline, and die. It is therefore reasonable to project senescence, and declining diameter is one way to do this. However, the ATA equation is arbitrary; its coefficients are not derived from empirical data. Trees of different species and in different settings will senesce at a wide variety of ages and under varied circumstances.

It is therefore more reasonable to leave the trunk area unadjusted and simply address factors relating to tree health and structure using the adjustment for condition (Chapter 5). This affords the appraiser sufficient flexibility to address the issue. Moreover, Chapter 5 addresses how the functional replacement cost method can be used to model the cost of a smaller-diameter plant that provides benefits equivalent to those of a larger plant being appraised.

Cross-Sectional Area

Trunk cross-sectional area can be calculated from either trunk diameter or circumference (Appendix 2). The more a tree's cross section deviates from a circle, the smaller its true area will be for a given circumference.

When calculating the cross-sectional area of a multi-stem tree, compute the cross-sectional areas first, then the sum of the areas (Figure 4.5). Computing the sum of the diameters and calculating cross-sectional area from the sum is incorrect and will overestimate total cross-sectional area.

Crown Area/Crown Volume

The crown is the upper part of a tree or shrub, including all the branches and foliage. Plant size can be expressed as a crown area or volume. Crown measurements may be used to calculate the two-dimensional area either vertically (average height by average width) or horizontally (the average of two representative measures of width, such as north–south or east–west).

The area within the drip line is calculated by measuring crown spread radius or diameter in two or more directions, as described above. When estimating the three-dimensional volume of the crown, the formulas applied vary according to its shape (Appendix 2).

Plant Placement

Plant placement has two aspects: the plant's location on the site and the influence of the site on plant growth.

Plant location may be described generally (e.g., two large trees on the north side of a home) or specifically (e.g., Global Positioning System [GPS] coordinates). Generalized placement information is necessary during early stages of an appraisal to identify the subject tree(s) and avoid confusion. More specific placement information may help to verify which plant or landscape item was appraised. For some appraisals, architectural drawings, GIS mapping, or GPS mapping showing plant placement may be necessary, especially on large sites. A quick site sketch may be helpful.

Plant placement information may include (Urban Tree Growth & Longevity Working Group 2016):

- location descriptions such as median plantings, street trees, or specific property locations;
- site maps, either sketches or formal drawings;
- reference objects that provide distance and orientation;
- GPS coordinates;
- identification tags;
- photographs illustrating the relationship to other site features; and
- unique benefits provided, such as screening or view enhancement.

Placement is important. First, plant appraisers consider the services provided by the plant that can be affected by location. In climates with warm summers, trees on the south and west sides of a home are more effective in reducing summer energy use than trees on the north and east. Plant placement will influence the effectiveness of a hedge for screening.

The second aspect of placement considers the impact of location on plant growth and potential functional limitations. In general, growth limitations occur when a plant's growth potential is constrained by the proximity to structures or activities. Common placement limitations include overhead utility lines, underground gas lines, nearby structures, pavement, and other plants.

Plant Function

Plants can provide shade, enhance views, create a focal point, and screen adjacent properties. Plants have many traditional uses as well as contemporary applications, such as green roofs, vertical gardens, rain gardens, and biowalls.

How effectively plants achieve their purpose (function) is generally evaluated by professional judgment. Programs such as i-Tree Eco, however, can quantify the value of benefits such as the removal of atmospheric contaminants or energy conservation. The function of plants can change over time as when an orchard or woodland is converted to greenspace in a residential neighborhood.

Plant Age

Knowing tree age gives the appraiser insight into potential longevity. Tree species have widely different life spans, whether in natural or landscape settings (Burns and Honkala 1990).

Depending on the assignment, appraisers may need to determine the actual or relative age of the plant and/or its useful life expectancy. Relative age is based on the development stage. Plants are often described as young, semimature, mature, or overmature (senescent).

Plant age can be determined from growth ring counts or installation records. The installation date can be estimated by examining planting invoices or site development records. Historic photographs may provide estimates of planting or development dates in the absence of other documentations (see Google Earth [https://www.google.com/earth/]).

In temperate climates, the counting of annual growth rings may be a reliable indicator of tree age and growth rates. Growth rings are apparent on cut stumps or trunk cross sections. For standing trees, increment borers can be used to extract a core of wood from the trunk to count rings (Maeglin 1979). If rings are indistinct, staining or fine sanding may improve visibility.

This procedure is less dependable in Mediterranean and tropical climates. In these locations, false growth rings can develop in response to changes in soil moisture level, temperature, light exposure, or other factors and may be difficult to distinguish from true rings. Under such conditions, estimating tree age by counting rings may be less reliable. For example, a Monterey cypress (*Hesperocyparis macrocarpa*) with a 42-inch (107-cm) diameter growing in San Francisco had more false rings than true rings. Without cross-dating this core with known references, tree age could only be estimated as 48 to 125 years.

Plant Condition

For woody plants, *condition* is a general term that incorporates health, structure, and form (Table 4.1). These three components are not strictly independent of one another and each is integral to the condition rating needed for analyzing depreciation (Figure 4.6) (Chapter 5). Condition (and each of its components) can be expressed in several ways: (1) in qualitative terms (excellent, good, fair, poor, very poor, dead), (2) on a scale of numbers (0, 1, 2, 3, 4, etc.), and (3) in percentages based on a scale of 0% (lowest) to 100% (highest) (Table 4.1). Many appraisers use percentage ratings because they fit well into depreciation ratings.

There are a wide range of options for assessing plant condition. For example, *Best Management Practices: Managing Trees During Construction* (Fite and Smiley 2016) suggests a method where health and structure are assessed on a scale of 1 to 15 and age on a scale from 1 to 10. *Best Management Practices: Tree Inventories* (Bond 2014) describes a scale for condition with four levels (dead, poor, fair, good).

Whatever method a plant appraiser chooses to rate condition, the method should be consistently applied and verifiable. A condition rating should reflect the species' characteristics and the plant's stage of development.

There is no best time to evaluate plant condition, particularly for deciduous trees. In winter, foliage is absent but overall tree structure is readily apparent. In summer, the opposite is true. Plant appraisers should be prepared to evaluate health, structure, and form at any time of the year.

Plant Health

Assessing plant health has traditionally considered factors based on a visual inspection such as vigor, foliage size and color, leaf density, presence or absence of pests (e.g., insects, disease, or parasites), twig growth rate, amount of twig or branch dieback, and wound closure (Table 4.1). The seriousness of the symptoms and their effect on plant health may only be fully understood when the cause is diagnosed. Plant health should be evaluated during the field inspection.

Many insect and disease problems can affect the utility or value of a plant. Of greatest concern are those that will significantly shorten the life of the tree. These can be vascular wilt diseases, root rot and decay, canker

Table 4.1 Assessment of plant condition considers health, structure, and form. Each may be described in rating categories that can be translated into a percent rating.

Rating category	Condition components			Percent rating
	Health	Structure	Form	
Excellent	High vigor and nearly perfect health with little or no twig dieback, discoloration, or defoliation.	Nearly ideal and free of defects.	Nearly ideal for the species. Generally symmetric. Consistent with the intended use.	81% to 100%
Good	Vigor is normal for the species. No significant damage due to diseases or pests. Any twig dieback, defoliation, or discoloration is minor.	Well-developed structure. Defects are minor and can be corrected.	Minor asymmetries/deviations from species norm. Mostly consistent with the intended use. Function and aesthetics are not compromised.	61% to 80%
Fair	Reduced vigor. Damage due to insects or diseases may be significant and associated with defoliation but is not likely to be fatal. Twig dieback, defoliation, discoloration, and/or dead branches may comprise up to 50% of the crown.	A single defect of a significant nature or multiple moderate defects. Defects are not practical to correct or would require multiple treatments over several years.	Major asymmetries/deviations from species norm and/or intended use. Function and/or aesthetics are compromised.	41% to 60%
Poor	Unhealthy and declining in appearance. Poor vigor. Low foliage density and poor foliage color are present. Potentially fatal pest infestation. Extensive twig and/or branch dieback.	A single serious defect or multiple significant defects. Recent change in tree orientation. Observed structural problems cannot be corrected. Failure may occur at any time.	Largely asymmetric/abnormal. Detracts from intended use and/or aesthetics to a significant degree.	21% to 40%
Very poor	Poor vigor. Appears to be dying and in the last stages of life. Little live foliage.	Single or multiple severe defects. Failure is probable or imminent.	Visually unappealing. Provides little or no function in the landscape.	6% to 20%
Dead				0% to 5%

diseases, boring insects, and many others. If a potentially fatal insect or disease affects the tree, the health component of the condition rating may be low or very low (Table 4.1).

Abiotic disorders include damages from chemicals such as herbicides, misapplied pesticides, air pollutants, heating fuel, natural gas, chlorine, or excessive fertilizers. Chemical damage occurs if a material is misapplied or escapes from its intended use, either accidentally or intentionally. Plant response depends on the chemical, the amount that was applied, the time it was applied, the weather, the sensitivity of the plant, and the health of the plant.

Plants affected by construction projects or improper planting procedures may decline over several years or decades. The damage may be due to physical injury to branches, trunk, or roots; soil compaction; changes in soil moisture level; girdling; or other injuries.

Figure 4.6 Examples of plant health, structure, and form in Monterey cypress. (a) Mature tree with typical form, good structure, and excellent health. (b) Tree in poor health but with normal structure and form. (c) Tree in good health and typical form but with compromised structure: crack between codominant stems. (d) Tree with good health and structure but with an asymmetric and one-sided crown.

Plant health ratings should be reduced if there is trunk damage or a significant loss of roots or branches. Because some species tolerate more damage than others, the amount of reduction will be based on species, age, and site conditions. If a vigorous, young tree were to lose 20% of its bark around its trunk circumference, the main effect would be visual. The effect on a mature tree is more severe. If 50% or more of the roots or branches are lost, or if more than 50% of the trunk circumference has bark loss, health would be rated as poor or very poor in most cases.

Health ratings need to consider the time frame of the appraisal. For retrospective appraisals, the appraiser may need to determine the cause of the injury, its extent, and the condition of the affected plant(s) before the damage. The pre-event health is often the starting point for determining any change. Determining responsibility usually is not the function of the plant appraiser, unless it is part of the assignment.

For current appraisals, the health ratings should be based on the current conditions, not future expectations. For example, the health rating of an American elm should not be downgraded due to its susceptibility to Dutch elm disease. It should only be downgraded if the tree actually has the disease. Features inherent within the genotype of the plant, such as susceptibility to Dutch elm disease, would be considered as a functional limitation, and if the pest is in the area, that is an external limitation (Chapter 5). For prospective appraisals, anticipated condition at the effective date of the valuation should be estimated.

Plant Structure

An appraiser visually examines the plant to develop a structure rating (Table 4.1). In addition to inspecting for defects and other structural problems, an appraiser should also look for response growth, i.e., woody tissue produced to strengthen areas of mechanical weakness (Table 4.2).

Features that reduce structural integrity are not independent of one another. Several factors may interact and reduce strength. A leaning tree may or may not be stable, but a leaning tree with a crack in the lower trunk is less stable and would receive a lower structure rating.

Appraisers should bear in mind that an examination for structural defects as part of an appraisal assignment does not constitute a tree risk assessment (Smiley et al. 2017). A plant appraiser may encounter a situation where a tree risk assessment is advisable but is not part of the assignment. Under such conditions, the appraiser should communicate to the client the need for such a service. Such communication should be clearly stated and permanently documented.

Foresters collect other information on product grade (quality), though this is technically the product of both timber size measurements (e.g., diameter, log length, and number of defects) and subjective observations (e.g., crook, sweep, rot, and branches).

Plant Form

Form describes the plant's habit, i.e., its shape or silhouette (Figure 4.7). Form develops from the interaction of genetics, management history, pruning system, and the environment. Plants of one species can be expected to have a similar form when grown in similar conditions. For example, open-grown eastern white pine, sweetgum, and coast redwood will have a pyramidal form. American elm will have a broad vase-shaped form. Weeping willow and weeping beech will have just that, a weeping form. Cultivars are genetically identical and will tend to have the exact same form. That said, form may change as the plant ages.

Form may be modified by pruning, and this should be considered when evaluating the plant. Pruning systems define the long-term form. Many trees are pruned with the intent to maintain the tree's natural shape, size, growth pattern, and adaptations. Other pruning systems include topiary, pollard, and espalier. A pollard can have excellent form even though it is not the "natural" shape of the plant.

Form is one of the key criteria in selecting plants for use in the landscape. A wise horticulturalist once observed, "The more natural the plant looks, the more beautiful it is." Successful function, aesthetic appeal, longevity, and intensity of maintenance are all influenced by plant form. At the same time, tree form is often manipulated to provide a special effect. Hedges and topiary are two examples.

Another aspect of tree form is its position relative to other trees in the landscape or forest stand, often categorized as crown class (Figure 4.8):

- dominant, i.e., standing above the canopy of neighboring trees
- codominant, i.e., at the same general level as the main canopy, receiving light from above but not from the sides

Table 4.2 Checklist for assessment of tree structure.

Indicators of decay	Cavities, fruiting bodies Wounds Cankers and galls Nesting holes Carpenter ants
Roots	Cavities, wounds Mounded soil Exposed or uplifted roots Change in grade (e.g., fill soil) Absence of flare Stem girdling roots Discolored, decayed, or dead roots
Trunk	Deviation from the vertical (lean, bow, sweep) Codominant or multiple attachments with or without included bark Cracks, seams, ribs Taper Dead tissue, decay, missing bark Cankers Abnormal flare (e.g., bottle butt)
Crown	Orientation (vertical, horizontal) Symmetry (one-sided, end weight, overextended) Codominant or multiple attachments with or without included bark Poor branch attachment Cracks, seams, ribs Broken, partially attached, or hanging branches Dieback and decline of branches or twigs Missing bark, decay, dead branches Cankers
Site factors	History of failure Drainage Recent changes General forest characteristics
Management history	Site change (construction, grading, exposure) Pruning (crown, roots) Structural support (cable, brace, prop, guy) Lightning protection

- intermediate, i.e., shorter in height than codominant trees and receiving light only from directly above
- suppressed, i.e., beneath the general level of the main canopy

Trees that have been suppressed or are in the intermediate class are not likely to have or develop good form. When evaluating plant form, the important questions are as follows:

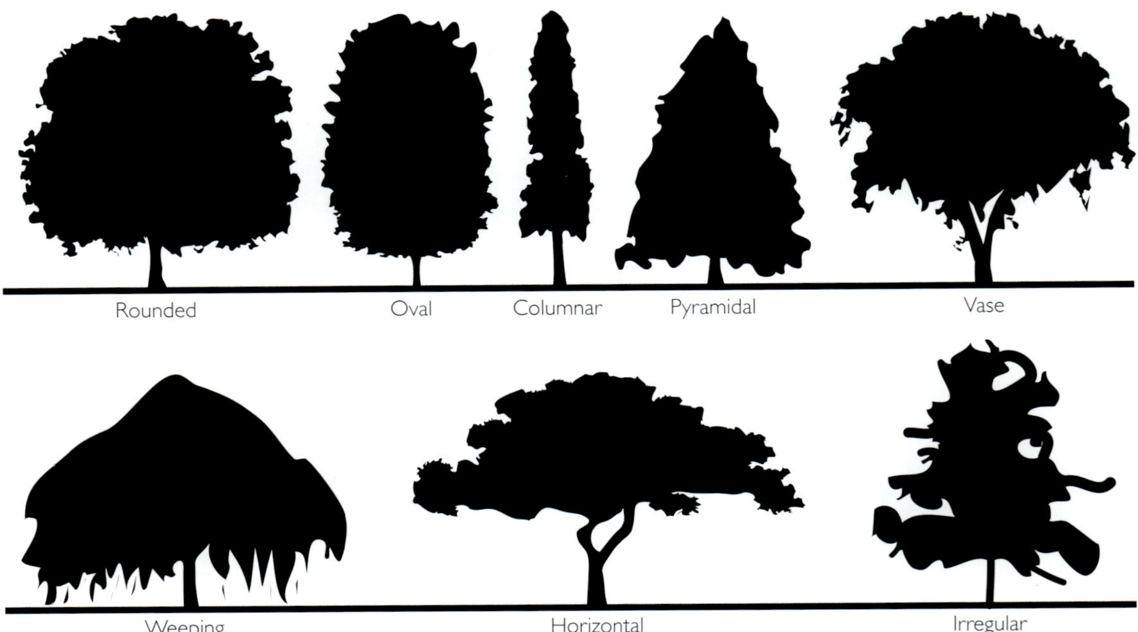

Figure 4.7 Examples of tree form. (Illustration by Tim Toland.)

Figure 4.8 Crown class. D (dominant), CO (codominant), I (intermediate), and SU (suppressed). (Illustration by Nelda Matheny.)

- Is the form of the plant consistent with the natural form of the species, growing conditions, management history, and stage of development?
- Has the plant's form been manipulated to achieve a desired function or appearance? Pruning can alter form in a way that enhances plant appearance and/or improves a specific function. Pruning can also alter form in a manner that detracts from appearance or function.
- What factors have caused an atypical form? Aside from management activities like pruning and applying growth regulators, plant placement may alter form. Growth of trees placed below overhead powerlines, close to structures, and near other plants may never achieve an ideal form.

Form should not be confused with structure (Figure 4.9). Because form focuses on the overall shape and silhouette of the crown, its assessment is largely based on aesthetic and functional considerations. Form can be

Figure 4.9 Although this tree does not have the classic natural form of the species, it offers significant aesthetic appeal.

judged as excellent even if a tree is not tall, straight, or symmetric, so long as it provides a high level of function and aesthetic appeal.

Combining ratings of plant health, structure, and form into the condition rating. The appraiser needs to combine the assessment of plant health, structure, and form into an overall condition rating. This can be accomplished in several ways. First, there is the intuitive option, whereby the appraiser can

Table 4.3 Using weighted averages to establish an overall condition rating of a Deodar cedar (*Cedrus deodara*) tree. Refer to Figure 4.10.

Component	Rating	Weighting	Product
Health	1.00	0.15	0.15
Structure	0.60	0.70	0.42
Form	0.40	0.15	0.06
Sum	2.0	1.0	0.63
Weighted Average Condition Rating (sum of product/sum of weighting)			**0.63 ÷ 1**
Weighted Average Condition Rating			**0.63**

use professional judgment to determine the overall condition rating, whether by qualitative, numeric, or percentage estimations.

Second, the appraiser can use the lowest individual rating to establish the overall condition rating. In this approach, if health was poor but structure and form were good, the overall rating would be poor.

A third approach employs a weighted average of the three components (Table 4.3; Figure 4.10). This process involves four steps. First, health, structure, and form are evaluated in decimal form and the results added together. Second, the appraiser considers whether one of these components is more important than any other and, if so, applies a weighting factor. Third, the ratings of health, structure, and form are multiplied by the weighting factor. Fourth, the product of the rating and weighting are added together and divided by the sum of the weighting.

For example, the Deodar cedar in Figure 4.10 was assessed with a health rating of 1.00 (100%), a structure rating of 0.60 (60%), and a form rating of 0.40 (40%). In the second step, the appraiser judged that the three factors were not equal in importance to the overall condition rating, with structure being more important than either health or form. The appraiser weighted the significance or importance of each individual component to overall condition: health, 0.15 (15%), structure, 0.70 (70%), and form, 0.15 (15%). In the third step, the rating of each component was multiplied by the weighting factor. Fourth, the product of each component is

Figure 4.10 Example of weighted averaging for condition. Deodar cedar (*Cedrus deodara*) located close to the foundation of a school building. The central leader had been lost. Multiple attachments arose at 18 feet (5.5 m). Several low branches swept upright but were bowed outward at the tips. Other branches were horizontal and extended over sidewalk and steps. The east side had been pruned several times to clear the building and roof, resulting in a one-sided crown. Tree health was excellent. Structure was fair. Form was compromised.

added together and divided by the sum of the weighting (0.15 + 0.42 + 0.06) ÷ 1.0. In this example, the result was a weighted average of 63%, equivalent to good condition (Table 4.1).

The CTLA does not recommend averaging the three component ratings, particularly where one component is very low while the other two are high. For the cypress trees in Figure 4.6, a tree with excellent form and structure that was dead could receive ratings of 100%, 100%, and 0%. The component average would be 200 divided by 3, or 67%. Such a rating would indicate the tree was in good condition even though it was dead.

There is no single best way to compute an overall condition rating, but the process should be thoughtful and credible, not arbitrary and mechanical.

Assessing Physical Damage to Landscape Plants

Estimating physical damage combines qualitative assessment and quantitative measurement. Type and extent of injury can be documented with measurements, descriptions, and photographs. After deciding if the plant is a complete or **partial loss**, appraisers may assess tree condition twice: (1) prior to the damaging event (retrospective) and (2) after the damaging event (current or prospective).

Damage assessment should consider the following questions:

- What plant parts are affected?
- Has plant health, structure, and/or form been compromised?
- What is the extent of the damage (e.g., percentage of trunk girdling, percentage of crown loss)?
- Is the damage short term (plant will recover) or permanent?
- Has plant longevity been reduced?
- Can the plant be restored to the predamage condition by pruning or other treatment?

Damage to the main trunk, scaffold limbs, or roots may not manifest for years. Tree health and structure are adversely affected in proportion to the extent of loss. The appraiser's task, then, is to assess the proportion of benefit loss. Generally, but not always, if half or more of the bark on the trunk, half or more of the root system, or half or more of the crown is lost on a mature tree, it is considered a total loss. Belowground injury is more difficult to assess than aboveground damage.

For example, a moving van runs into a 60-year-old Japanese maple (*Acer palmatum*). Approximately 15% of the branches are broken. The tree's appearance and symmetry are damaged. Tree health, structure, and potential longevity, however, are not. Aside from removing the damaged branches, no additional treatment may be needed. The amount of damage can be quantified in the rating of the tree's form when using the depreciated cost method.

The determination of proportional loss and its long-term significance is a matter of professional judgment, which includes consideration of species characteristics as well as the benefits provided by the plant or tree before and after the damaging event.

Management History

Information regarding the history of landscape plantings and maintenance may be important. Data collection considerations for understanding horticultural practices may include the following:

- determining if plants are planted, native, or naturalized;
- assessing planting density (planted or natural);
- reviewing past fertilization, irrigation (including water quality), pruning, and pest management (insects, disease, weeds);
- site disturbance (from cultivation or construction); and
- soil properties and management.

Off-Site Research

In addition to on-site data collection, information may also be acquired through off-site research. As always, the type of data needed will depend upon the appraisal problem. One of the most common examples of this type of data is cost information for the purchase and installation of landscape plants. If the appraisal problem involves market value, information on real estate markets and recent transactions is necessary. Literature searches can document plant information (such as life span) and apply appropriate research findings. Relevant ordinances and regulations may be found.

If highest and best use (HBU) of a property is needed, focus research on the stability of the residential neighborhoods. If owner turnover is generally low and homes are typically renovated rather than torn down, this is an indication of stability. Historic neighborhoods may have regulations to limit activities or dictate aesthetic considerations (including appropriate plant species). In contrast, transitional neighborhoods can be subject to redevelopment pressures. In these neighborhoods, plans may call for significant lot reconfiguration or changes in zoning, building density, or road design.

Commercial and industrial districts can be highly variable in their dynamics as well, ranging from stable use to those likely to be redeveloped. These changes can include new commercial or industrial uses, as well as wholesale redevelopment into multi- and mixed-use neighborhoods.

Rural areas can also be highly variable. Development pressures can be significant as agricultural land sold for development can be highly lucrative in some markets, and highway improvement encourages further sprawl from urban centers. In these cases, HBU may change in the short- to midterm. Countering this in some regions are land preservation initiatives that seek to preserve rural character and protect the environment.

Appraisers may need to be aware of the development dynamics in their region. When there is a high potential for land use change, the appraiser should consider contacting local planning and regulatory agencies to identify whether there are any short- to midterm plans for an area or neighborhood.

CHAPTER 5

The Cost Approach

CHAPTER OUTLINE

Overview

The cost approach produces a cost estimate for repairing, replacing, or restoring the utility of the item. The cost approach is often applied to damaged or destroyed items, tree inventories, preconstruction bonds, and insurance claims. The principle of substitution (Chapter 2) is the foundation of the cost approach and can be critical to assessing the reasonableness of the conclusions (Appraisal Institute 2015: p. 225).

Within the cost approach, there are three methods and three techniques which may be used to develop a **basic cost**. Selecting a specific method and technique depends on the assignment and the judgment of the appraiser. Figure 5.1 depicts the general process of the cost approach.

Appraisers can estimate either **direct cost** or **extrapolated cost**. Direct costs reflect actual or estimated costs for labor, materials, equipment, and supplies used to install, treat, or maintain the landscape item. Extrapolated cost begins with the cost of the largest commonly available nursery-grown tree and then extrapolates that cost to the size of the appraised tree. Plant appraisers commonly apply two techniques for extrapolating cost: the **trunk formula technique** (**TFT**) and the **cost compounding technique** (**CCT**, also called *cost forwarding*).

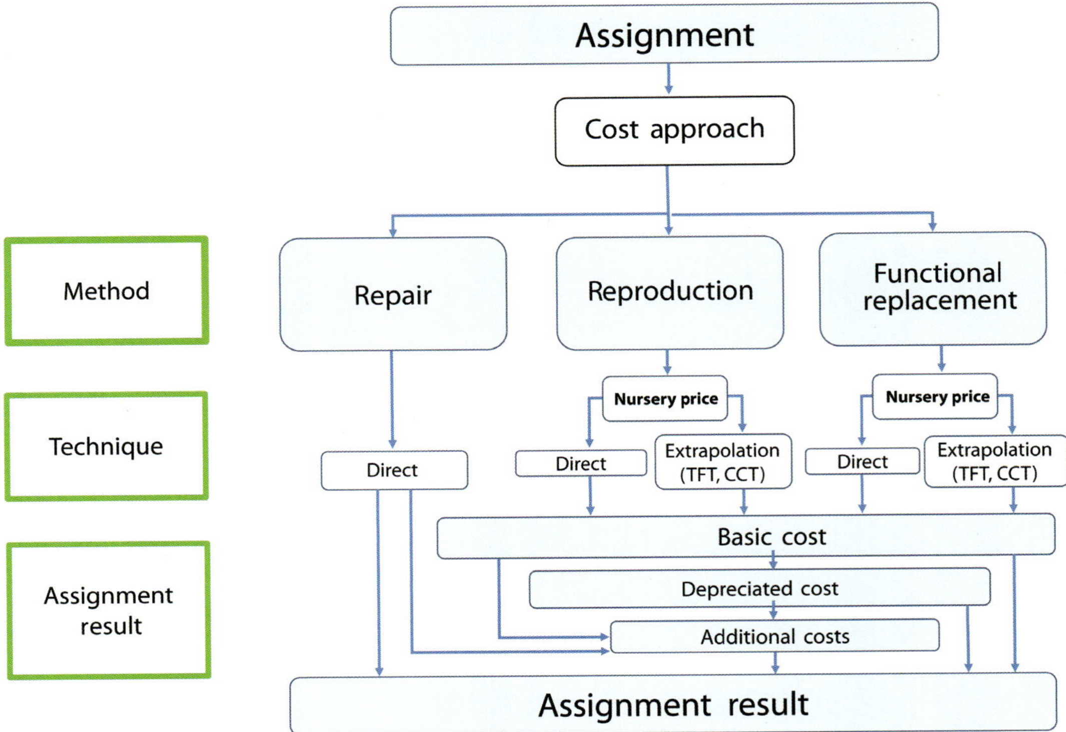

Figure 5.1 Flow chart of the cost approach.

Because the appraised tree or item is usually neither new nor perfect, it may be necessary to account for certain suboptimum characteristics. Such a deduction or depreciation considers physical deterioration, functional limitations, and external limitations. A depreciated cost estimate is the result of applying depreciation to a basic cost.

In some appraisals, there are additional costs beyond the replacement plant that need to be considered. These may include the cost of removing the damaged plant, installing the new plant, site clean-up, and plant maintenance during the establishment period.

The assignment result can be the basic cost; the depreciated cost; the basic or depreciated cost plus additional costs; or a reconciliation among several approaches, methods, or techniques. Where landscape components are appraised, the assignment result should not be considered market value unless it is tied to a defined market, as with real estate market value. The appraisal report should clearly define the assignment result and cite the source for the definition.

NEW NAMES

In this edition of the *Guide*, the names of several methods and techniques were changed in order to align plant appraisal terminology with that used by the larger appraisal profession and by the legal profession:

- Cost of cure is now functional replacement cost.
- Replacement cost is now reproduction cost.
- Trunk formula method is now trunk formula technique.

Methods for Estimating Cost

The appraiser's selection of a particular method and technique is a function of the appraisal problem, context of the assignment, direction provided by the client, and constraints imposed by law (cases, statutes, ordinances). The repair cost method addresses the cost to repair damages or mitigate further loss. The reproduction method estimates a cost of an exact replica of the item being appraised. Functional replacement cost estimates the cost to restore benefits.

Repair Cost

Repair cost is used when there is damage to a plant or other landscape feature and the assignment focuses on correcting the damage or mitigating further losses. Application requires two assumptions: (1) the item will remain in place, and (2) it will continue to provide benefits similar to those it provided prior to damage. Unlike the other two methods, repair cost does not usually estimate the cost to return the plant or landscape item to its predamage condition or utility. For example, a large-diameter broken branch cannot be reattached, but an arborist can be hired to prune the damaged branch. The repairs in this case involve mitigating future damage (e.g., the tearing of bark if it fails), promoting wound closure, and removing the debris.

Reproduction Cost

Reproduction cost is commonly used where a landscape item has been destroyed, removed, or significantly damaged. This methodology is widely used for tree inventories, preconstruction bonding, and some insurance claims.

Reproduction cost is the cost to replicate or duplicate the item being appraised. Generally, this means estimating the cost of replacing the landscape item with one that is close to identical (i.e., the same species or brand, size, shape, and condition) and thereby providing most or all of the characteristics and benefits of the original. When depreciation is applied to a reproduction cost, the result is termed a **depreciated reproduction cost**.

Functional Replacement Cost

Functional replacement is the cost of substitute items that provide equivalent utility, benefits, or function, rather than the cost to produce an exact replica. The principle of substitution is the foundation of the functional replacement methodology: a prudent person would not spend more to purchase an item or restore its benefits than the cost of a substitute item that produced similar benefits. The functional replacement cost method is used in many of the same situations in which the reproduction cost method is used, but it provides a valuation where the emphasis is on restoring benefits as opposed to duplicating a landscape feature.

For instance, a reasonable replacement for a damaged medium-size tree that provides shade or screening may be a similarly sized tree of a different species, a smaller tree of the same species, several smaller trees, a landscape structure, or some combination of these.

Functional replacement cost starts with assessing the utility of, or benefits provided by, the item. The appraiser then develops a plan that would produce a landscape that is functionally similar to the original landscape and is reasonable and appropriate for the site. It may be a similar plant, a group of plants, or an alternative landscaping item.

The functional replacement cost may be depreciated when the original item was not ideal. However, one of the advantages of functional replacement is that it reduces or eliminates the need for certain types of depreciation. This is particularly true when considering landscape items that do not add a value or utility, a condition known as superadequacy.

Techniques for Estimating Cost

The direct cost technique (DCT), trunk formula technique (TFT), and cost compounding technique (CCT) are techniques applied within the cost approach. The DCT estimates actual costs to install or repair a landscape item. The TFT and CCT are **extrapolation** techniques that take the cost of a nursery plant and proportionally increase it to infer the cost of a larger plant.

SUPERADEQUACY

As described in Chapter 2, superadequacy is an excess in the capacity or quality of a structure or structural component which does not add value or functional utility to an object or property. A tree that is too large for the growing space or intended function may be considered superadequate in that a smaller tree may provide the same level of benefits (see Figure 5.2).

For example, a strategically placed shade tree may reduce heating and cooling costs. If a 40-foot (12.2-m) tree provides the same energy savings as a 60-foot tree, the 60-foot (18.3-m) tree may be considered to be superadequate. Alternatively, a large-growing tree beneath overhead utility lines creates a conflict that may be considered superadequate due to size.

Superadequacy is relevant to functional replacement because the method focuses on utility and benefits. Either a direct or extrapolated cost may be estimated based on the size of the tree necessary to replace the benefit or utility. Such a tree may be smaller than the subject tree(s).

When applying the reproduction method, superadequacy may or may not be relevant. If the appraisal problem calls for estimating a cost to reproduce the item, any superadequacy is irrelevant.

Figure 5.2 Two types of superadequacy. (a) Number. Five pines were installed in a space better suited for a single tree. (b) Size. The aesthetic and energy conservation benefits provided by this large cypress could also be provided by a much smaller specimen.

With all techniques, basic cost estimates reflect an ideal condition. This estimate may be depreciated for suboptimum conditions, and/or increased to account for additional costs associated with clean-up, planting, and future maintenance.

Direct Cost Technique (DCT)

The direct cost technique is preferred when plants equivalent to the size of plants being appraised are commonly available. The direct cost technique totals the costs of plants, services, or other materials needed to repair, reproduce, or functionally replace the item. This is an estimate of actual cost, realistic in terms of its scope, availability of materials and contractor services, and feasibility.

This process starts with a plan for repair or replacement. The appraiser then estimates and tabulates all costs associated with the project (see worksheet example at the end of the chapter). DCT may even be useful for appraising large trees where the appraiser can find a local, large-tree moving company that can procure and install a tree similar to the damaged or removed tree.

Only those treatments directly related to the damage should be included. For example, pruning broken branches caused by a truck collision is appropriate, but removing all dead branches throughout the canopy would improve on the predamage condition of the tree, so it would not be appropriate. Depending on the assignment, depreciation may be applied. Additional costs for clean-up, delivery, planting, or future maintenance may be added when reasonable and appropriate.

Extrapolated Costs

When appraising trees or shrubs that are larger than commonly available from a local nursery, the appraiser can extrapolate from the cost of a nursery plant using the trunk formula technique (TFT) or cost compounding technique (CCT).

For common landscape applications, use the largest commonly available nursery plant as the basis for these calculations. Regional Plant Appraisal Committees (RPAC) can provide guidance on selecting the diameter of nursery trees to use with these techniques (see Appendix 4). For forestry or woodland settings, it is generally more appropriate to use the cost of seedlings rather than larger nursery trees to develop the basic cost.

The nursery plant cost is the price a landscape professional would pay for the plant at the nursery. It should not include delivery, planting, staking, or any other services. However, those and other costs, including site clean-up and maintenance during the establishment period, may be included as additional costs.

STRENGTHS AND LIMITATIONS OF THE TRUNK FORMULA TECHNIQUE

Strengths

- It is a cost-effective and simple way to appraise large trees.
- It is relatively easy to calculate unit cost from RPACs, online data, or nursery catalogs.
- It has a long history of use and acceptance by arborists, landscape professionals, and legal communities.

Limitations

- It is based on the assumption that the cost of a nursery tree can be reliably scaled to the cost of a large tree. In many situations, this lacks empirical basis.
- Cost estimates may be greatly out of proportion to the value of the land and other property improvements, or to what people would actually pay for a replacement tree.
- Application is generally limited to residential and urban landscape settings.

Trunk Formula Technique (TFT)

The trunk formula technique extrapolates the costs to purchase the largest commonly available nursery plant to the size of the plant being appraised (Figure 5.3). An underlying inference is that the cost to acquire a large plant is directly proportional to the unit cost of the nursery plant.

Unit cost is the cost per square inch (or cm^2) of trunk area, feet (or m) of trunk height, square feet (or m^2) of canopy projection, or cubic feet (or m^3) of crown volume. Arborists commonly use the price per square inch of trunk cross-sectional area measured at 4.5 feet (Chapter 4).

To apply the TFT using trunk diameter, compute the cross-sectional area of the subject plant then multiply it by the unit price (see Appendix 2). For example, a 3-inch-diameter nursery tree has a cross-sectional area

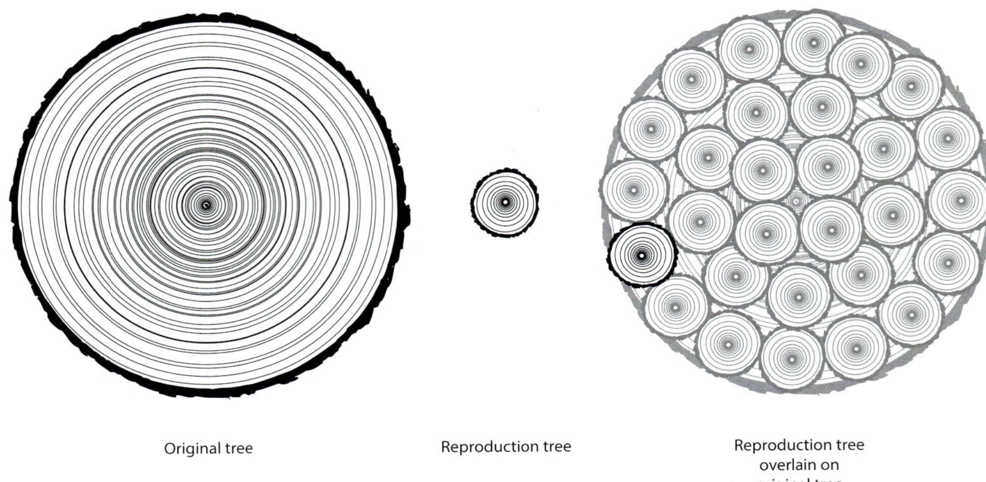

Figure 5.3 The trunk formula technique (TFT). This extrapolates the cost to produce a tree of a specific size in a nursery to that of a subject tree.

of 7.07 square inches (3.14 × 1.5 in × 1.5 in) and costs \$500 to purchase. The unit cost is calculated as \$500 ÷ 7 in^2 = \$70.77/in^2. If the subject tree had a dbh of 20 inches, its cross-sectional area is 314 square inches (3.14 × 10 in × 10 in) and its basic cost is \$22,222 (314 in^2 × \$70.77/in^2).

The CTLA recommends extrapolating the cost of purchasing the largest commonly available nursery tree because that cost is tangible and related to larger trees. The CTLA advises against extrapolating planting and additional costs when applying TFT because of the weak relationship between these costs for a nursery tree and for a much larger tree.

When possible, the nursery tree should be the same species and cultivar as the appraised tree. For functional replacement, extrapolation may be applied to different species or to a smaller tree that will replace the benefits of the appraised tree. One difference between reproduction and functional replacement is that calculations can be based on the estimated diameter of the functional replacement tree(s) rather than the actual diameter of the specimen being appraised. For example, if you are appraising a 40-inch-dbh (101.6-cm) tree, and a 30-inch-dbh (76.2-cm) tree would provide the same function and benefits in the landscape, use the 30-inch-dbh as the starting point for the TFT. Alternatively, several small specimens of similar or mixed species may provide a more cost-effective way to produce the same (or even superior) benefits. This is true where crown size is more important than overall trunk diameter.

A similar process is used when the landscape is excessively vegetated (i.e., there is superadequacy). If there are ten 25-inch (64-cm) trees in the front yard of a suburban, residential quarter-acre lot, the loss of one tree may have no impact on the overall benefits, so the value can be appraised at zero or each tree can be depreciated to reflect superadequacy. Excessive tree count can detract from both urban landscapes and woodlots.

i-TREE COMPENSATORY (STRUCTURAL) VALUE

The i-Tree suite of urban forest management tools (www.itreetools.org) uses a variation of the trunk formula technique to estimate "compensatory" value (Nowak et al. 2002). Specifically, the calculation employs the location values based on land use, as described in the eighth edition (1992) of the *Guide*. Therefore, the results are not directly comparable to those using the trunk formula technique described in this edition.

i-Tree Eco also uses a different system to monetize the environmental and ecological benefits provided by trees (see Chapter 6).

In some insurance cases, the extrapolated cost may be the assignment result. In other cases, this basic cost may need to be depreciated and/or have additional costs added to it. The additional cost may include tree removal, site clean-up, planting, and/or future maintenance.

Where future maintenance is required, proposed costs can be projected, adjusted for increasing costs over time if necessary, and then converted to present value using an appropriate discount rate (Chapter 6).

Palm and Shrub Appraisal Using the Trunk Formula Technique

When a large palm cannot be acquired, moved, and installed, an extrapolation technique can be applied. The unit pricing for palms is usually based on plant height. The height measurement will vary with region and species. The typical unit is either height of clear trunk or total height.

To apply the TFT to palms, ascertain which height increment is used in the region as the basis for pricing at local nurseries. Then ascertain the unit price from local nurseries or the Regional Plant Appraisal Committee.

The nursery cost can vary greatly with species and location, so the unit cost applied should be from the same or similar species. Multiply the height of the appraised palm by the unit cost to determine the basic cost. Depreciation and additional cost can be applied when appropriate.

Height-based or crown-volume-based pricing also applies to shrubs in many areas. The procedure is similar to that described for palms but is based on crown projection or crown volume.

STRENGTHS AND LIMITATIONS OF THE COST COMPOUNDING TECHNIQUE

Strengths

- Cost estimates are based on documentable tree cost and interest rates, which are defendable, empirical, based on biological and financial facts, and consistent with the principle of substitution.
- Calculations are easily performed.
- It works best where the years to parity can be reliably estimated.
- It can generally be applied in both urban landscape and wildland settings.
- Results reflect actual investor behavior in certain forestry applications.

Limitations

- Results are sensitive to years-to-parity estimates and interest rates.
- Arborists may need to seek professional advice when selecting interest rates.

Estimates of tree value may be disproportionate to the value of the land and other property improvements.

Cost Compounding Technique (CCT)

Cost compounding (syn. *cost forwarding*) is an extrapolation technique that relates the cost of money over a specific time period to tree growth. The appraiser estimates the time required for a new planting to either attain equivalent size or provide similar benefits or utility as the subject tree and then compounds installed cost for that time period using an appropriate interest rate (Figure 5.4).

There are three primary inputs to the CCT: (1) the installed cost of the nursery tree (present installed cost, *PC*); (2) the time in years it will take to reach equivalent size or utility (n); and (3) the appropriate compound interest rate (i). The formula for cost compounding is

$$CC = PC \times (1 + i)^n$$

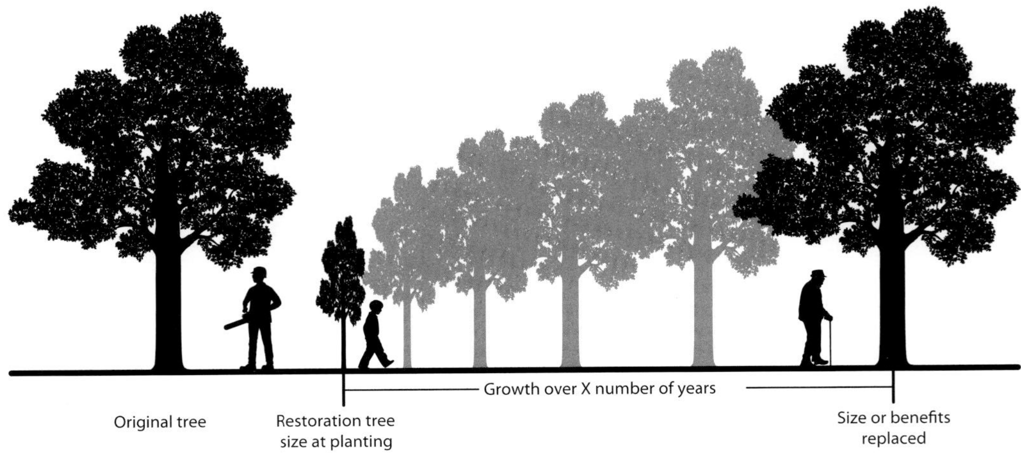

Figure 5.4 Cost compounding extrapolates the cost of tree replacement to a time in the future when the size or benefits of the appraised tree will be attained.

where CC = compounded cost, PC = present installed cost of the nursery tree, n = years for the new tree to reach parity with the appraised tree, and i = interest rate.

The present installed cost (PC) of the tree is the installed cost of the largest commonly available nursery plant. It includes the cost for the nursery plant, transportation, planting, and aftercare. Additional costs include site preparation, post-planting weed control, mulch, irrigation, and fertilization. The appraiser should follow the practices common to local landscapers or timber growers.

Unlike the TFT, the CCT extrapolates all establishment costs, not just the cost of the tree at the nursery. This is because the economic principle behind CCT involves an investor foregoing alternative use of capital when investing in the establishment of a tree.

The number of years (n) to attain equivalent size or benefits is sometimes called years to **parity**. Size parity may be based on tree diameter, crown spread, or crown volume. The appraiser should apply credible growth rates. This can be done by consulting local experts or by examining nearby trees of the same species. Data may be obtained from direct measurements, planting records, crown size comparisons from inspection of photographic images, or increment core or stump growth-ring counts.

Selecting an interest rate (i) to use in the compounding formula is a key step and can have a substantial impact on the extrapolated cost estimate. The rate should be relevant to the type of property being appraised as well as reasonable and defensible. As in the functional replacement method, the concept of interest rate selection is based in the principle of substitution. A prudent investor would not pay more (or accept less) for the expected benefits of the subject tree than for an equivalent substitute investment that produces cash-equivalent benefits.

Appraisers may research the rate of return of an agricultural mutual fund or tree planting or forestry investments as an appropriate fit. For residential properties, current mortgage rates or prime plus 2.0% may be applicable. For forestry or other commercial real estate investments, the rate is often determined using the client's commonly required rate of return for such investments.

Timber investors and appraisers commonly apply CCT to develop acquisition offers and estimate market value. In commercial forest applications, CCT involves projecting the costs of seedlings, site preparation, and planting to the age of the subject trees. It is most useful for young trees that have not reached merchantable size. CCT produces results that are closely tied to market value because the appraiser starts with the costs of seedlings, not large nursery stock. Applying CCT to older, mature trees often results in unrealistic value estimates. It may cost $300 per acre to plant 600 trees, or 50 cents per tree. Here, the extrapolated cost per tree (at 5%) is only $19 for a single tree that is 75 years old.

Table 5.1 Annual compound interest factor.

Years	Annual interest rate 2%	3%	4%	5%	6%	7%	8%	9%
1	1.02	1.03	1.04	1.05	1.06	1.07	1.08	1.09
3	1.06	1.09	1.12	1.16	1.19	1.23	1.26	1.30
5	1.10	1.16	1.22	1.28	1.34	1.40	1.47	1.54
10	1.22	1.34	1.48	1.63	1.79	1.97	2.16	2.37
25	1.64	2.09	2.67	3.39	4.29	5.43	6.85	8.62
50	2.69	4.38	7.11	11.47	18.42	29.46	46.90	74.36
75	4.42	9.18	18.95	38.83	79.06	159.88	321.20	641.19

Compound interest = $(1 + i)^n$, where i = interest rate and ^{n} = years.

CALCULATING THE UNIT COST

The unit cost is the price per trunk cross-sectional area (dollars per square inch) of the largest commonly available nursery-grown tree. In both the eighth and ninth editions of the *Guide*, the appraiser and RPACs were given the flexibility to use either the wholesale, retail, or installed price as the basis for determining the unit cost.

In this edition of the *Guide*, the CTLA recommends using the price that a landscape professional would pay a wholesale nursery grower to purchase the plant as the basis for unit cost.

Table 5.1 presents the compounded cost (*CC*) of a dollar (*PC*) invested for *n* years at various interest rates (*i*). It illustrates how sensitive calculations are to the choice of interest rate and number of years it takes to reach parity.

As with any cost approach technique, it may be necessary to depreciate the basic cost to account for condition, functional limitations, and external limitations (as described later in this chapter). Additional costs for clean-up, planting, and maintenance may need to be added.

Depreciation

Depreciation is the monetary expression of suboptimum factors (Chapter 2). Appraisers use depreciation to account for the differences between the cost of the new or ideal item and the item being appraised, which typically has some lower level of quality due to less than ideal features, its placement, or the site that it occupies.

General appraisal practice holds that depreciation is the combination of three factors: (1) physical deterioration, (2) functional obsolescence, and (3) external obsolescence. Physical deterioration reflects plant condition and considers structural integrity, health, and form (Chapter 4). Functional limitations are factors associated with the property or the tree itself that limit future plant development (Table 5.2). External limitations are factors outside the property and the control of the tree owner that affect life expectancy, structure, health, or form.

Table 5.2 Common functional limitations.

General category	Potential limitation
Placement	Available growing space Overhead utilities Underground utilities Nearby structures, sidewalks, roads
Superadequacy	Excess size Excess plant density
Soil	Volume available for growth Chemistry (lack or excess of specific mineral elements)
Plant genetics	Disease and insect susceptibility Fruit and litter Thorns Invasiveness Failure pattern Tolerance to construction activities Root or basal sprouting Allergenicity Soil and water requirements (pH, alkalinity)
Water	Quality Irrigation adequacy/excess Drainage Water table

DEPRECIATION FACTORS

Previous editions of the *Guide* described three depreciation factors: species, condition, and location. To avoid issues with double counting and to present a framework that is more consistent with the general appraisal community, this edition presents three categories of depreciation: physical deterioration (condition), functional limitations, and external limitations.

Physical Deterioration (Condition)

The components of condition are health, structural integrity, and form (Table 4.1). A tree that is ideal with regards to these components is rated at 100%. A dead tree is typically rated at or near 0%. The final condition rating combines ratings of the three components. Chapter 4 provides more detail about evaluating condition.

Functional Limitations

Functional limitations are factors associated with the interaction of a tree and its planting site and will affect plant condition, limit development, or reduce utility of the plant within the foreseeable future. These factors

Table 5.3 Examples of functional limitation ratings.*

Situation	Pruning system	Example rating	Notes
SPECIMEN TREE			
	Any	100%	No limits to growing space.
LARGE MATURING TREE			
Near property line	Any	10% to 90%	Based on proportion of canopy growing into neighboring property and disrupting site use.
Under powerline	Headed/round over	5%	—
Under powerline	Through-trimmed	30% to 70%	—
Adjacent to powerline	Side-trimmed	30% to 70%	—
Under powerline	Hedge	75% to 100%	Tree is managed to control height.
Between curb and sidewalk; adequate tree lawn	Any	25% to 75%	Presence or potential for root/pavement conflict.
Between curb and sidewalk; narrow tree lawn	Any	10% to 40%	High potential for root/pavement conflict.
SMALL MATURING TREE			
Under powerline	Natural	90% to 100%	—
Adequate tree lawn	Natural	75% to 100%	Presence or potential for root/pavement conflict.
Narrow tree lawn	Natural	50% to 75%	Presence or potential for root/pavement conflict.
SHRUBS			
Close to house foundation, sidewalk, driveway, etc.	Natural	10% to 75%	Based on proximity to foundation and need for clearance pruning and/or height control.
Near property line	Hedge	80% to 100%	Based on proximity to property line and need for clearance pruning.
SPECIES WITH FRUIT AND/OR LITTER			
Residential lawn	Natural	80% to 100%	Depending on use and maintenance of lawn.
Sidewalk, parking, bench, or other use area	Natural	10% to 25%	Based on degree of disruption of site use and function.
Landscape bed	Natural	80% to 100%	—
INVASIVE SPECIES			
	Any	0% to 20%	Based on state or regional listing and potential for disrupting native vegetation.
TREE OR SHRUB SUSCEPTIBLE TO LETHAL PEST IN THIS AREA			
	Any	10% to 30%	

* Percentage ratings are for illustrative purposes only and intended to be used with sound appraiser judgment.

DEPRECIATING FOR SPECIES

Depreciation for species has historically been included in the *Guide*. The species factor was originally included because the basic cost was aimed to the ideal species, i.e., one rated as 100%. Where species were less than ideal, the species factor allowed depreciation to be applied. Species ratings commingled genetic factors, response to growing conditions, business practice, and personal appreciation of aesthetics.

In this edition of the *Guide*, functional limitations incorporate a species genotype with a specific location, allowing for a more defined assessment of performance.

include site conditions, placement, and genetic limitations (Table 5.2). If the species and site present significant restrictions to growth, performance, and function, then the depreciation should be significant (Tables 5.2 and 5.3). Functional limitations are considered either incurable or requiring repeated or costly treatments to mitigate. For example, a young tree that will be large at maturity that is located under powerlines, within a narrow tree lawn, or susceptible to a lethal pest will receive significant depreciation because these factors will decrease life expectancy; lead to a deterioration in health, structural integrity, or form; or require repeated treatments to mitigate the condition.

External Limitations

External limitations are factors that are outside of the property, out of the control of the property owner, and that will affect plant condition, limit development, or reduce plant utility within the foreseeable future. These factors include legal restrictions that limit the development of the plant and environmental factors that affect long-term health and life expectancy of the plant. Examples of external limitations include the following:

- laws, ordinances, and easements that grant to other parties the authority to prune or remove vegetation impinging on powerlines, obstructing views, or blocking solar access;
- water use limitations, restrictions on irrigation;
- competing infrastructure (utilities);
- the presence of serious pests in the area; and
- changing climate zones.

For example, a tree on one property screens an existing solar collector on an adjacent property in an area where solar access is protected by local regulations. The tree must be pruned or removed to maintain solar access. This is a limitation on tree development that is outside the control of the tree owner. For this reason, depreciation for external limitation should be applied. The amount of depreciation may reflect the amount of canopy that will remain and/or the loss of benefit and utility.

Another example of external limitation is an airport glide-path easement that restricts the height of trees grown on adjacent property. The owner of the easement has the right to top or remove the trees.

External limitations for more intangible factors like low neighborhood property values are more challenging to assess and may be relevant only where the assignment result is market value.

If the appraiser finds no external limitations present, the depreciation rating should be 100%.

Applying Depreciation

When applying depreciation to a basic cost, the appraiser assigns a multiplier ranging from 0% to 100% to each of the depreciation categories: condition, functional limitations, and external limitations (see Table 5.4). The basic cost is multiplied by each of the three categories to estimate the depreciated cost. The resulting depreciated cost may be the assignment result.

Table 5.4 Summary of depreciation factors and suggested ratings.

Condition (overall assessment of health, structure, and form)	**Functional limitations** (assessment of species-site interaction)	**External limitations** (assessment of outside factors that influence plant success)
Excellent (81% to 100%)	No impact (81% to 100%)	No impact (81% to 100%)
Good (61% to 80%)	Minor impact (61% to 80%)	Minor impact (61% to 80%)
Fair (41% to 60%)	Moderate impact (41% to 60%)	Moderate impact (41% to 60%)
Poor (21% to 40%)	Severe impact (21% to 40%)	Severe impact (21% to 40%)
Very poor (6% to 20%)	Extreme impact (0% to 20%)	Extreme impact (0% to 20%)
Dead (0% to 5%)		

Appraisers may find that some features fit into more than one depreciation category. For example, overhead electrical wires can be either a functional limitation or an external limitation. In this case, t'he appraiser should depreciate in only one category.

Additional Costs

Basic or depreciated costs may need to be adjusted to account for costs associated with clean-up, planting, and future maintenance. These additional costs can include removal of the damaged branches, tree, or debris; preparation of the site; installing the new plant; site clean-up; irrigating the new plantings; treating pests; and/or otherwise restoring the tree or landscape to predamage condition or as close to those conditions as is practical.

Estimates for these services and materials may come from proposals prepared by contractors, other professionals, or the appraiser, if qualified and not conflicted. It may be appropriate to obtain several estimates and select the most reasonable and appropriate for the situation.

Additional costs should be projected as far into the future as necessary, based on a reasonable and practical assessment by the appraiser. This is often limited to the establishment period. If the assignment is to appraise the additional costs for the first year after replacement planting, then it should include the cost of one year of care. If it is for ten years after planting, then it should include the anticipated costs for the next ten years of care, discounted at an appropriate interest rate (Appendix 5).

All treatments recommended should be consistent with industry standards and best practices. Only those treatments directly related to the damage should be included. For example, pruning broken branches caused by a truck collision is appropriate, but removing all dead branches throughout the canopy would improve on the predamage condition of the tree. Root, soil, and other treatments may be appropriate if they will aid in recovery, provided the treatments and costs are neither uncommon nor excessive.

When large plants or landscape features are involved, or when site conditions dictate the use of heavy equipment or specialized techniques, it may be appropriate to include resulting cost of collateral damage.

As with all appraisal data, additional costs should be reasonable and estimated from the perspective of the effective date of the valuation.

Summary

The cost approach produces a cost estimate for repairing, reproducing, or restoring the utility of the item. Three methods and three techniques can be used to develop a basic cost. Selecting a specific method and technique depends on the appraisal problem and the judgment of the appraiser. Appraisers can estimate either direct cost or extrapolated cost. Direct costs reflect actual or estimated costs for labor, materials, equipment, and supplies used to install, treat, or maintain the landscape item. Extrapolated cost begins with the cost of the largest commonly available nursery-grown plant and then extrapolates that cost to the size of the appraised tree. Plant appraisers commonly apply two techniques for extrapolating cost: the trunk formula technique (TFT) and the cost compounding technique (CCT).

Because the appraised tree or item is usually neither new nor perfect, it may be necessary to account for certain suboptimum characteristics. This deduction or depreciation considers physical deterioration, functional limitations, and external limitations. In some appraisals, there are additional costs beyond the replacement plant that need to be considered.

The assignment result can be the basic cost; the depreciated cost; the basic or depreciated cost plus additional costs; or a reconciliation among several approaches, methods, or techniques. As described in Chapter 2, where landscape components are appraised, the assignment result should not be considered market value unless it is tied to the market value of the real estate of which it is a part.

Examples of Depreciation

Example 1. Monterey Cypress (*Hesperocyparis macrocarpa*) Tree Adjacent to Powerline

Condition

- Structure: excellent, 90%
- Health: excellent, 90%
- Form: fair, since the tree is one-sided, 70%

Functional limitations

- It is an appropriate species-site match with adequate growing space, except for the proximity of the powerline.
- Clearance pruning is required for the powerlines over the street and over the sidewalk.
- Species is susceptible to cypress canker, which is common in the area.

External limitations

- The powerline easement is outside the control of the tree owner.
- The clearance requirements over street and sidewalk are outside the control of the property owner.

Depreciation

Each of the depreciation factors reflects the same situation: the proximity of a large-growing tree to energized conductors. The powerlines were present before the tree was installed. Were it not for the powerlines, the cypress would be in excellent condition. The tree is close enough to the powerlines and vigorous enough that continued trimming is required.

The plant appraiser must take care not to depreciate for the same situation among the three categories. Here, the appraiser's judgment is to account for the electrical lines as a functional, rather than an external limitation. The final depreciation could appropriately be as follows:

Condition: 90%

Functional limitations: 50% (see Table 5.3)

External limitations: 100%

The total depreciation factor may be computed as 90% × 50% × 100% = 45%, representing total depreciation of 55%.

Example 2. Coast Live Oak (*Quercus agrifolia*) with Crack Between Codominant Stems

Condition

- Structure: very poor, 10%
- Health: excellent, 90%
- Form: good, 70%

The low structure rating reflects an imminent likelihood of failure due to the crack.

Functional limitations

- The buried root collar may increase susceptibility to disease and tree failure.
- The species-site match is excellent for this area.

External limitations

- View easement is in the area, but the tree will need to grow 30 feet (9.14 m) higher to be affected, so it is not a concern in the short or moderate term.

Depreciation

There is an excellent match of site and species and an absence of external limitations, but it is a structurally unsound tree. The final depreciation could appropriately be as follows:

Condition: 10%

Functional limitations: 80%

External limitations: 100%

The depreciation factor may be computed as 10% × 80% × 100% = 8%, representing total depreciation of 92%.

Example 3. Dying Monterey Pine (*Pinus radiata*) in Park

Condition

The structure and form are typical of mature trees of this species: lost central leader, some weak branch attachments. Canopy raising over time resulted in a live crown ratio of 30%. Red turpentine (*Dendroctonus valens*) and five-spined engraver (*Ips paraconfusus*) beetles are present in the tree.

- Structure: fair, 50%
- Health: very poor, 5%
- Form: fair, 60%

Functional limitations

- The species is susceptible to numerous species of bark beetles and borers and there are no practical treatment options.
- The species is susceptible to pine pitch canker disease (*Fusarium circinatum*) and there are no treatment options.

External limitations

- Property owner does not permit pesticide application.

Depreciation

There is an excellent match of site and species. Tree health is compromised by an untreatable insect infestation. The final depreciation could appropriately be as follows:

Condition: 5%

Functional limitations: 90%

External limitations: 90%

The depreciation factor may be computed as 5% × 90% × 90% = 4%, representing total depreciation of 96%.

Example 4. London Plane (*Platanus × hispanica*) Street Tree Beneath Overhead Powerlines

Condition

Tree was pruned as a pollard early in its development. Pollarding was replaced by topping. The crown has been permitted to grow on either side of the wires.

- Structure: fair, 50%
- Health: fair, 50%
- Form: poor, 30%

Options for combining the condition ratings:

1. Intuitive: fair, 45%
2. Lowest rating (form): poor, 30%
3. Weighted average: 34%
   a) Weighting: structure, 0.40; health, 0.30; form 0.30
   b) Weighted average: (50% × 0.40) + (50% × 0.30) + (30% × 0.30) =
   20% + 15% + 9% = 44% ÷ 1 = 44%

Functional limitations

- The presence of overhead utility lines limits performance.
- The species performs well in the region.
- The species tolerates topping and reduction pruning. It has a low rate of branch failure.
- The planting space is marginally adequate; some trees are likely to lift the adjacent sidewalk.
- The species is infected by anthracnose in the spring and powdery mildew in late summer/early fall.

External limitations

- The powerline easement is outside the control of the tree owner.
- The clearance requirements over street and sidewalk are outside the control of the property owner.

Depreciation

Tree structure, health, and form are less than ideal. The species performs well in this type of site. Disease problems are a nuisance but do not threaten long-term viability. The final depreciation could appropriately be as follows:

Condition: 45%

Functional limitations: 50%

External limitations: 80%

The depreciation factor may be computed as 45% × 50% × 80% = 18%, representing total depreciation of 82%.

Example 5. Pollarded Deodar Cedar (*Cedrus deodara*) in Residential Area of Large City

Condition

The tree has been pollarded for many years to manage its size and form.

- Structure: excellent, 90%
- Health: good, 70%
- Form: excellent, 90%

Options for combining the condition ratings:

1. Intuitive: excellent, 85%
2. Lowest rating (health): good, 70%
3. Weighted average: 35%
   a) Weighting: structure, 0.80; health, 0.10; form, 0.10
   b) Weighted average: (90% × 0.80) + (70% × 0.10) + (90% × 0.10) =
   72% + 7% + 9% = 88% ÷ 1.0 = 88%

Functional limitations

- The species performs well in the region.
- There is limited rooting area, but with this pruning system, it does not have a major impact.

External limitations

- Residents on the second floor have a view easement. The tree has to be maintained below 30 feet (9.14 m).

Depreciation

The pruning system manages a tree that would otherwise be too large for this site. The final depreciation could appropriately be as follows:

Condition: 82%

Functional limitations: 90%

External limitations: 80%

The depreciation factor may be computed as 82% × 90% × 80% = 59%, representing total depreciation of 41%.

Examples Using the Cost Approach

Example 1. Repair Method Using the Direct Cost Technique

A car drives into the front yard of Mrs. Renee's property, leaving ruts in the grass, destroying a concrete birdbath, and knocking over a pear tree before slamming into an 18-inch-dbh (46-cm) oak. The car owner's insurance company hires an appraiser to determine the cost to repair the landscape. The appraiser first develops a plan to repair the damage:

1. Clean up debris from the displaced soil, birdbath, and damaged trees.
2. Fill in the ruts with topsoil.
3. Cover the fill soil with sod of the same grass species that is in the rest of the yard.
4. Replace the birdbath.
5. Pull the pear tree to a vertical angle, guy, and anchor.
6. Remove the loose damaged bark from the oak.
7. Irrigate the new sod for one month.
8. Irrigate the trees for two growing seasons.
9. Treat the oak tree as follows:
    - Apply an insecticide to prevent insect attack.
    - Apply a fungicide to reduce the risk of *Phytophthora* infection.
    - Fertilize according to the soil analysis to promote health and growth.
    - Mulch to conserve soil moisture.

A cost is then estimated for each item on the list:

Ex 1 Appraisal estimate to repair damage.

Task	Details	Cost
1	Clean-up. 2 people, 1.5 hours @ $75 = $225.00, dump charge $10.	$235
2	Fill ruts. 2 people, 1.75 hours @ $75 = $262.50. Soil cost $70.	$333
3	Sod laying. 2 people, 1 hour @ $75 = $150. Sod cost $125.	$275
4	Replace birdbath.	$125
5	Guy the pear tree, pick up and install birdbath. 6 hours @ $75.	$450
6	Treat oak wound. 1 person, 1 hour @ $175.	$175
7	Temporary irrigation: hoses, timer, sprinklers.	$160
8	Water.	$15
9	Pest management (one plant health care visit).	$275
10	Soil analysis and fertilization.	$200
11	Mulch.	$150
Total		$2,393
Round to		**$2,400**

Example 2. Mr. Butler's Dead Hedge

Mr. Butler has a hedge of mature arborvitae (*Thuja occidentalis*) plants along one property line. In an attempt to provide insect control, Mr. Butler's neighbor Ms. Peabody inadvertently applies an herbicide to the hedge, resulting in the death of the arborvitae. Mr. Butler is anxious to replace the hedge as soon as possible. Its main benefit was to screen Mr. Butler's view of Ms. Peabody's home. Mr. Butler retains a plant appraiser to estimate the cost of replacing the hedge in support of an insurance claim.

Prior to the damage, the height of the hedge was 15 feet (4.57 m), with the most effective screening in the lower 8 feet (2.44 m). The plants were growing at a rate of about one foot per year. They were healthy except for an infestation of bagworms, which resulted in some twig dieback.

Example 2a. Functional Replacement Method Using the Direct Cost Technique

The plant appraiser and Mr. Butler discuss three options that would restore the utility of the hedge as soon as possible:

1. Install a new hedge composed of 10-foot-tall (3.05-m) plants.
2. Install a 6-foot-tall (1.83-m) fence.
3. Install a 6-foot wood lattice fence planted with vines.

Option One: New Hedge

1. Remove dead and dying trees and resulting debris. 8 hours @ $50.	$400
2. Purchase and install 10-foot-tall arborvitae. 10 @ $250.	$2,500
3. Purchase and install equipment for temporary irrigation.	$150
4. Hire PHC contractor for one growing season. 6 visits @ $100.	$600
Total	**$3,650**

Option Two: New Fence

1. Remove dead and dying trees and resulting debris. 8 hours @ $50.	$400
2. Install 30 feet (9.14 m) of 6-foot colored vinyl solid fence.	$1,800
Total	**$2,200**

Option Three: New Fence Plus Vines

1. Remove dead and dying trees and resulting debris. 8 hours @ $50.	$400
2. Install 30 feet of 6-foot colored vinyl lath fencing.	$1,500
3. Install 20 3-gallon (13.56-L) pot climbing hydrangeas.	$1,800
4. Purchase and install equipment for temporary irrigation.	$150
Total	**$3,850**

The three options are presented to Mr. Butler without reconciliation.

Example 2b. Reproduction Method Using the Direct Cost Technique

The plant appraiser and Mr. Butler discuss reproduction of the hedge as it existed the day before Ms. Peabody applied the herbicide. The plant appraiser determines that 15-foot-tall (4.57-m) arborvitae plants are available locally.

Basic Cost

Purchase ten 15-foot tall arborvitae. 10 @ $400.	$4,000

Depreciated Cost

The plant appraiser depreciates for condition, because the hedge was not in perfect condition at the time of the loss.

Condition (health, structure, form): 85%

Functional limitations: 100%

External limitations: 100%

Subtotal, depreciated cost ($4,000 × 0.85 × 1.00 × 1.00)	**$3,400**

Additional Costs

Remove dead and dying trees and resulting debris. 8 hours @ $50.	$400
Installation cost. 10 trees @ $100.	$1,000
Purchase and install equipment for temporary irrigation.	$150
Hire contractor to provide PHC/IPM services. 6 visits @ $100.	$600
Subtotal, additional costs	**$2,150**
The total, depreciated reproduction cost ($3,400 + $2,150)	$5,550
Round to	**$5,600**

Example 2c. Reproduction Method Using the Trunk Formula Technique

The plant appraiser also considers the trunk formula technique. Trees 10 feet tall (3.05 m) are available in the nursery, but trees 15 feet tall are not. The plant appraiser assumes a linear increase in the purchase cost as plants increase in height.

Basic Cost

Extrapolate the cost of 10-foot-tall arborvitae to 15 feet tall. Each 10-foot-tall tree costs $150, so the unit cost of a 10-foot-tall tree is $15 per foot of height ($150 ÷ 15 feet). Extrapolating the unit cost to a 15-foot-tree yields a basic cost of $225 per 15-foot tree.

10 replacement trees @ $225 per tree.	$2,250

Depreciated Cost

The plant appraiser depreciates for condition, because the hedge was not in perfect condition at the time of the loss.

Condition (health, structure, form): 85%

Functional limitations: 100%

External limitations: 100%

Subtotal, depreciated cost ($2,250 × 0.85 × 1.00 × 1.00)	**$1,912.50**

Additional Costs

Remove dead and dying trees and resulting debris. 8 hours @ $50.	$400
Installation cost. 10 trees @ $100.	$1,000
Purchase and install equipment for temporary irrigation.	$150
Hire contractor to provide PHC/IPM services. 6 visits @ $100.	$600

Subtotal, additional costs	**$2,150**
Total, depreciated reproduction cost $1912.50 + $2,150)	$4,062.50
Round to	**$4,100**

Example 2d. Reproduction Method Using the Cost Compounding Technique

The plant appraiser also uses the cost compounding technique to estimate a reproduction cost.

Basic Cost

10 replacement trees @ $225.	$2,250
Installation cost. 10 trees @ $100.	$1,000
Purchase and install equipment for temporary irrigation.	$150
Hire contractor to provide PHC/IPM services. 6 visits @ $100.	$600
Basic cost	**$4,000**

Compounded Cost

$$CC = PC \times (1 + i)^n$$

where CC = compounded cost, PC = current establishment costs, i = interest rate, and n = years to parity.

Adjustable rate mortgage	2.80%
Years to parity	5
Compounding factor	1.14806
Basic cost compounded	**$4,592**

Depreciated Cost

Condition (health, structure, form): 85%

Functional limitations: 100%

External limitations: 100%

Subtotal, depreciated basic cost ($4,592 × 0.85 × 1.00 × 1.00)	**$3,903**

Additional Costs

Remove dead and dying trees and resulting debris. 8 hours @ $50.	$400
Subtotal, additional costs	**$400**
Total, depreciated compounded cost ($3,903 + $400)	$4,303
Round to	**$4,300**

Example 3. Illegal Tree Removal: Reproduction Cost Using the Trunk Formula Technique

Hansom Home Builders removes a tree in the public right-of-way as part of a residential property teardown and rebuild project. The village ordinance does not allow for the removal of healthy street trees to accommodate private property construction. The appraiser is hired by the construction company to estimate a reproduction cost, which will be used to establish the fine imposed by the Village.

The village ordinance calls for using the trunk formula technique for trees over 6 inches (15.24 cm) in diameter. The Village provided tree data from their inventory, which was updated three years earlier. The inventory identified the tree as 13-inch-dbh (33.02-cm) bur oak (*Quercus macrocarpa*). Tree condition was very good. The tree was located between the sidewalk and curb. There were no overhead powerlines or other external limitations.

Basic Cost

Cross-sectional area. (3.14 × 6.5 × 6.5) | 132.7 in^2
Unit cost of the largest commonly available nursery tree | \$58/in^2
Basic cost (132.7 in^2 × \$58/in^2) | **\$7,695**

Depreciated Cost

Condition: 90%

Functional limitations: 85%

External limitations: 100%

Total, depreciated cost (\$7,696 × 0.90 × 0.85 × 1.00) | \$5,886

Round to | **\$5,900**

Example 4. Ms. Thomas's Red Oak Killed by Lightning Strike

A large (44-inch-dbh or 111.76-cm) red oak (*Quercus rubra*) in Ms. Thomas's small front yard in Athens, Georgia, U.S., is killed by a lightning strike. Ms. Thomas wants to determine the cost of the tree so that she can submit a claim to her home-owners insurance company.

The tree was in good health and had been well managed. The main stem formed two co-dominant trunks at 10 feet (3.05 m). Energized conductors were present on one side of the tree. Bacterial leaf scorch disease is common in the area. Two other mature trees are present in the front yard. Other red oaks in the neighborhood are 24 to 30 inches (76.2 cm) dbh.

Ms. Thomas calls three local nurseries to obtain prices on their red oaks and establish the unit cost.

The plant appraiser has worked with the third nursery in the past and determines to use their costs. The principle of substitution might otherwise argue for using the lowest estimate (Nursery 1, \$40.74), but in this case the appraiser selects a higher estimate (Nursey 3, \$44.56) because of its superior tree quality and reputation for excellent customer service.

Example 4a. Reproduction Method Using the Trunk Formula Technique

Ex 4 Unit cost of red oaks.

Nursery	Caliper (in)	Cross-sectional area (in^2)	Purchase cost	Unit cost
1	5	19.6	\$800	\$40.74
2	4	12.6	\$700	\$55.70
3	5	19.6	\$875	\$44.56

Basic Cost

Cross-sectional area of 44-inch tree. (3.14 × 22 × 22) | 1,519.8 in^2
Unit cost of the largest commonly available nursery tree | \$44.56/in^2
Subtotal, basic cost (1,519.8 in^2 × \$44.56/in^2) | **\$67,726**

Depreciated Cost

Condition: 80%

Functional limitations: 75%

External limitations: 80%

Subtotal, depreciated cost ($67,726 × 0.80 × 0.75 × 0.80) $32,508

Additional Costs

Remove dead tree. Grind the stump to 6 inches (15.24 cm) below grade. Fill hole with soil.	$3,000
Planting.	$300
Temporary irrigation. One season of PHC.	$600
Subtotal, additional costs	**$3,900**
Total, reproduction cost estimate ($32,564.20 + $3,900)	$36,408
Round to	**$36,400**

Example 4b. Functional Replacement Method Using Trunk Formula Technique: Superadequate Situation

Based on observations of other trees in the area, the appraiser determines that a 24-inch tree would provide the same benefits as the 44-inch (111.76 cm) tree.

Basic Cost

Cross-sectional area of 24-inch (106.68 cm) tree (3.14 × 12 × 12)	452.2 in^2
Unit cost of the largest commonly available nursery tree	$44.56/$in^2$
Subtotal, basic cost (452.2 in^2 × $44.56/$in^2$)	**$20,160**

Additional Costs

Remove dead tree. Grind the stump to 6 inches below grade. Fill hole with soil.	$3,000
Planting.	$300
Temporary irrigation. One season of PHC.	$600
Subtotal, additional costs	**$3,900**
Total functional replacement cost ($20,160 + $3,900)	$24,060
Round to	**$24,100**

Example 4c. Reproduction Method Using the Cost Compounding Technique

The appraiser estimates the tree's age by counting 68 growth rings visible in the tree stump. If the trunk diameter is 44 inches and the bark thickness is 1 inch, then the diameter of the wood is 42 inches (106.7 cm). The average annual radial growth is 0.62 inches (1.58 cm) per year.

The trunk diameter of the replacement tree is 5 inches (12.5 cm), measured at 12 inches (30.48 cm) above grade, and its dbh is 4 inches (10.16 cm). The difference between the dbh of the replacement tree and the subject tree is 40 inches (101.6 cm). Based on annual radial growth of 1.24 inches (0.62 × 2), it will take the replacement tree 32 years to reach the diameter of the subject tree. Note: This assumes that the replacement tree will grow at the same rate as the subject tree. If there is evidence of retarded growth in the subject tree, due to past competition or other factors, it may be reasonable to project a lower age to parity.

Basic Cost

Replacement tree.	$875
Installation.	$300
Hire contractor to provide PHC/IPM services. 6 visits @ $100.	$600
Basic cost	**$1,775**

Compounded Cost

$CC = PC \times (1 + i)^n$

where CC = compounded cost, PC = current establishment costs, i = interest rate, and n = years to parity.

30-year fixed mortgage rate	5.0%
Years to parity	32
Compounding factor	4.765
Basic cost compounded	**$8,458**

Depreciated Cost

Condition (health, structure, form): 80%	
Functional limitation: 75%	
External limitations: 80%	
Subtotal, depreciated compounded cost ($8,458 × 0.80 × 0.75 × 0.80)	**$4,060**

Additional Costs

Remove dead tree. Grind the stump to 6 inches (15.24 cm) below grade. Fill hole with soil.	$3,000
Subtotal, additional costs	**$3,000**
Total, depreciated compounded cost	$7,060
Round to	**$7,100**

Example 4d. Estimating a Market Value Using the Trunk Formula Technique and Hedonic Regression Research

Ms. Thomas's insurance company requests an estimate of the loss of value in her property due to the loss of her 44-inch (111.76-cm) red oak. The appraiser decides to do this by combining the TFT cost estimate with the results of hedonic regression sales comparison research (Chapter 7).

Looking at Table 7.1, the appraiser decides that the hedonic regression analysis research conducted by Anderson and Cordell (1988) is the most appropriate to apply, because there are several trees in the landscape and they are in the same area where the research was conducted. This research shows that front-yard trees typically contribute 3.5% to 4.5% of the property value.

The appraisal starts with a depreciated, extrapolated reproduction cost estimate using the trunk formula technique. Since there are three trees in the front yard, the appraiser needs to estimate the cost of each tree, then compare that result to the sales comparison regression research estimate.

The other two trees in Ms. Thomas's front yard are a 20-inch-dbh (50.8-cm) willow oak and a 15-inch-dbh (38.1-cm) white oak. The appraiser estimates the depreciated reproduction cost of the three trees as follows:

Basic Cost for Red Oak

44-inch-dbh tree has a 1,520 in^2 cross section. ($A = r^2\pi = 22 \times 22 \times 3.14 = 1{,}520$)

Unit cost of the largest commonly available nursery tree is \$44.56/in^2

1,520 in^2 × \$44.56/in^2 = \$66,869

Depreciated Cost

Condition: 80%

- Health: good
- Structure: codominant stem
- Form: no issues

Functional limitations: overhead powerlines result in minor side-trimming and asymmetry; bacterial leaf scorch susceptibility, 75%

External limitations: bacterial leaf scorch in the area, 80%

Depreciated cost = basic cost \$66,880 × condition 0.80 × functional limitations 0.75 × external limitations 0.80 = \$32,097

Round to \$32,100

Basic Cost for Willow Oak

20-inch-dbh (50.8-cm) tree has 314 in^2 cross section. (10 × 10 × 3.14 = 314)

Unit cost of the largest commonly available nursery tree is \$38/in^2

314 in^2 × \$38/in^2 = \$11,932

Depreciated Cost

Condition: 90%

- Health: good
- Structure: no issues
- Form: no issues

Functional limitations: close to house, will need routine clearance pruning, 85%

External limitations: none known, 100%

Depreciated cost = basic cost \$11,932 × condition 0.90 × functional limitations 0.85 × external limitations 1.00 = \$9,127.98

Round to \$9,100

Basic Cost for White Oak

15-inch-dbh (38.1-cm) tree has 177 in^2 cross section. (7.5 × 7.5 × 3.14 = 176.6)
Unit cost of the largest commonly available nursery tree is \$48/in^2
177 in^2 × \$48/in^2 = \$8,478

Depreciated Cost

Condition: 80%

- Health: excellent
- Structure: codominant stems
- Form: no issues beyond codominant stems

Functional limitations: no limitations, 100%

External limitations: none known, 100%

Depreciated cost = basic cost \$8,496 × condition 0.80 × functional limitations 1.00 × external limitations 1.00 = \$6,782

Round to \$6,800

Total Cost

The total reproduction cost of all three front yard trees is \$48,000 (\$32,100 + \$9,100 + \$6,800).

It would cost \$300 to plant a replacement sapling and \$600 to provide temporary irrigation and one season of PHC/IPM. However, these two costs are not included in the analysis because they are implicit in the 3.5% to 4.5% range of contributory value indicated by hedonic regression research.

Market Value

The appraiser collects the following indications of market value for the property:

- Realtor.com: \$253,548
- Zillow.com: \$299,623
- Municipal property tax assessment: \$317,600

The appraiser selects $299,626 as the most reasonable estimate of current market value. The average contribution to market value for the trees in the front yard is 4.0% of the total property value, based on the hedonic regression research. Therefore, the subject trees contribute an estimated value of $299,623 × 0.04 = $11,985, which rounds to $12,000. Since this is for all three trees, the contribution of each tree needs to be estimated.

The contributory market value of the oak tree is $8,025. The additional cost to remove the dead tree, grind the stump, and fill the hole with soil is $3,000. Therefore, the total cost is $11,025, rounded to $11,000.

Example 4e. Estimating a Market Value Using TFT and Hedonic Regression Research

Ex 4e Estimating a Market Value using TFT and Hedonic Regression Research

Tree	TFT cost	% of total cost	Multiply by $12,000
Red oak	$32,100	67%	$8,025
Willow oak	$9,100	19%	$2,275
White oak	$6,800	14%	$1,700
TOTALS	$48,000	100%	$12,000
Total property	$299,623		
Percentage	4%		
Tree value	$11,985		
Round to	$12,000		
Contributory market value			$8,025
Additional cost			$3,000
Total cost			$11,025
Rounded			**$11,000**

Example 5. Partial Loss Using Reproduction Method with Trunk Formula Technique

The city arborist calls a tree appraiser to determine the monetary damages to a street tree that was struck by an out-of-control pickup truck. The appraiser's site examination finds that the tree in question is a 26-inch-dbh (66-cm) sugar maple. Bark is missing from a 28-inch-wide (71.1-cm) section of its circumference that extends from 6 inches to 5 feet (15.2 cm to 1.5 m) above the ground in the shape of a football.

Every year, the appraiser monitors the value of local nursery trees, so the appraiser knows that the largest commonly available nursery sugar maple has a 4-inch (10.2-cm) dbh and costs $1,350. The appraiser determines that it is possible to use the TFT before and after the damage incident to determine the loss.

Basic Cost

A 26-inch sugar maple has a 531 square inch cross-sectional area ($A = r^2\pi = 13 \times 13 \times 3.14 = 530.7$). The 4-inch nursery tree has a 12.56 square inch cross-sectional area ($2 \times 2 \times 3.14 = 12.56$) and costs $1,350. The unit price of the nursery tree is $107.48 per square inch ($1,350 ÷ 12.56 in^2). The extrapolated cost of the appraised tree is $57,038 (530.7 in^2 × $107.48/in^2), rounded to $57,000.

Partial Loss

The circumference of a 26-inch (66.04-cm) tree is 82 inches (208.28 cm) (26 in × 3.14 = 81.64 in). The appraised tree lost 28 inches (71.12 cm) of bark width, which is 34% of the tree circumference (28 in ÷ 82 in = 0.34). Based on this, the appraiser judges that the overall condition is poor and warrants a 30% rating. This produces a depreciated cost of $57,000 × 0.30 = $17,100, rounded to $17,000. Therefore, the estimated loss is the predamage cost of $57,000 less the postdamage depreciated cost of $17,000, or $40,000.

Example 6. Reproduction Cost Using the Trunk Formula Technique for a Palm

With palms, the TFT is typically based on plant height. The height is multiplied by the unit cost to compute the basic cost. This example involves appraising a 30-foot (9.15-m) *Phoenix dactylifera* palm, which is in good condition with some nutrient deficiency apparent on the older fronds. Some dead fronds remain attached. The palm is in an area where Texas Phoenix palm decline has been found. There are no placement limitations. Specimens of this size are not readily available in local nurseries, but shorter palms are available. The contractor price for a 10-foot (3.05-m) *Phoenix dactylifera* palm is $320.

Basic Cost

The palm being appraised is a 30-foot tall *Phoenix* palm. The largest commonly available nursery tree is 10 feet tall and costs $320. The unit cost of the largest commonly available nursery tree is $32 per foot. For 30 feet that amounts to $960.

Depreciated Cost

Condition: 85%

- Health: good
- Structure: dead fronds
- Form: normal

Functional limitations: Texas Phoenix palm decline susceptibility, 90%

External limitations: Texas Phoenix palm decline is in the area, 80%

Depreciated cost = basic cost $960 × condition 0.85 × functional limitations 0.90 × external limitations 0.80 = $587.52

Round to $590

Example 7. Functional Replacement Method Using Trunk Formula Technique for a Superadequate Landscape

A fire damages Mrs. Spice's landscape. At the advice of her lawyer, she hires an appraiser to estimate the monetary loss in the landscape. Her intent is to secure compensation for the damages from the neighbor who started the fire while burning trash on a windy day.

The appraiser examines the landscape and finds 12 dead 14- to 18-inch-dbh (35.6-cm to 45.72-cm) pines and 16 dead azaleas in the front yard of her house, which is on a quarter-acre lot in a middle-class suburban residential neighborhood. The azaleas are 3 feet (0.91 m) tall with a crown-spread diameter of 5 feet (1.52 m). The appraiser discovers that the pine trees were planted much closer together than on other properties in the neighborhood, and the overall benefits that the subject trees provide to the landscape could be provided just as well by only six trees.

The appraiser finds that the local nursery has 3-inch-dbh (7.62-cm) pine trees available in 48-inch (121.9-cm) boxes. The nursery also has azaleas with a 2-foot-tall (0.61-m) by 2.5-foot-diameter (76.2-cm) crown spread in 3.5-gallon (13.25-L) containers. The appraiser's functional replacement plan is as follows:

1. Clean up the site; remove dead and dying pines and azaleas.
2. Install 6 new pine trees and extrapolate their cost to the current size.
3. Install 16 tall azaleas and extrapolate their cost to the current size.
4. Provide 3 years of maintenance.

Basic Cost

Replacement pine trees in a 48-inch (121.9-cm) box cost $350 each, 7 square inches in cross section.

1.5 × 1.5 × 3.14 = 7.1

Round to 7

$350/7 in^2 = $50/in^2

Current trees have an average dbh of 16 inches (40.6 cm), so the cross-sectional area is 201 in^2 (8 × 8 × 3.14 = 201).

The extrapolated average cost of each tree using the TFT is 201 in^2 × $50/in^2 = $10,050, the estimated replacement cost of 6 trees (not 12 because of superadequacy) = 6 × $10,050 = $60,300.

Replacement azaleas are in 3.5-gallon (13.25-L) pots @ $100 each.

The crown volume of the replacement azaleas is calculated as a cylinder (height × radius^2 × π) = 2 ft × (2.5 ft/2)^2 × 3.14 = 9.81 ft^3 of crown volume. Therefore, the replacement cost is $100/9.81 ft^3 = $10.19 per cubic foot of crown volume.

The crown volume of the original plants was 3 ft (0.91 m), so 3 ft x (5/2)^2 x 3.14 = 58.88 ft^3.

The appraised cost of each azalea is $10.19 × 58.88 ft^3 = $600 each.

The appraised cost of all the azaleas is $600 × 16 plants = $9,600.

Additional Costs

Remove dead trees and shrubs.	$8,500
Install all plants.	$1,800
Maintenance for 3 years, including temporary irrigation and PHC/IPM.	$1,200
Total additional cost	$11,500
Total functional replacement cost:	
Trees $60,300 + shrubs $9,600 + additional $11,500	$81,400

Round to $81,000

Note that 3 years of maintenance costs a total of $1,200, or $400 per year. A more accurate estimate of this cost can be computed as the present value of three annual payments of $400. If the home mortgage rate is 5.0%, the present value of annual maintenance costs, each occurring at the beginning of the year, can be computed using Formula 2 from Appendix 5, Figure A5.1, as follows:

Year 1 = $400 (no discount for Year 1)

Year 2 = 400 ÷ 1.05^1 = $381

Year 3 = 400 ÷ 1.05^2 = $363

Total present value of future costs = $1,144

In this case, additional costs total $8,500 + $1,800 + $1,144 = $11,444. Total functional replacement cost is therefore $60,300 + $9,600 + $11,444 = $81,344, rounded to $81,000. This is a case where more precision does not produce materially different results. However, if the maintenance costs had been large or occurred over a long period of time, it would warrant the additional present value computation.

Repair Method
Direct Cost Technique

Client name ______________________________ Date __________ Case # __________

Phone ______________________________ E-mail ______________________________

Address ____

Subject tree

Species ____

Trunk diameter* (D) ________________ @ ______________

Damage description

____

____

____

Repair plan

____

____

____

Repair items

1. Cleanup ______________________________ $________
2. Wound repair ______________________________ $________
3. Pruning ______________________________ $________
4. Support system ______________________________ $________
5. Irrigation ______________________________ $________
6. Mulch ______________________________ $________
7. Turf ______________________________ $________
8. Shrubs ______________________________ $________
9. Other plantings ______________________________ $________
10. Soil ______________________________ $________
11. Hardscape ______________________________ $________
12. Debris removal ______________________________ $________
13. Aftercare ______________________________ $________
14. Other ______________________________ $________

Total repair cost[‡] (sum lines 1 through 14) $________

Rounded $________

* Diameter and cross-sectional area may be replaced with plant area, volume, or height as appropriate.
‡ Apply depreciation if it is appropriate for the assignment.

Reproduction Method
Trunk Formula Technique

Client name ______________________________ Date __________ Case # __________

Phone ______________________________ E-mail ______________________________

Address ______________________________

Subject tree

Species ______________________________

1. Trunk diameter* (D) ______________ @ ______________
2. Cross-sectional area (line 1)^2 × 0.7854 __________ in^2
3. Condition rating __________ %

   Health ______________________________

   Structure ______________________________

   Form ______________________________
4. Functional limitations ______________________________ __________ %
5. External limitations ______________________________ __________ %

Replacement tree

Species ______________________________

6. Trunk diameter* (D) ______________ @ ______________
7. Cross-sectional area (line 6)^2 × 0.7854 __________ in^2
8. Replacement tree cost Source: ______________________________ $__________

Calculations

9. Unit tree cost (line 8 / line 7 or RPAC) $__________
10. Basic reproduction cost (line 2 × line 9) $__________
11. Depreciated reproduction cost[‡] (line 10 × line 3 × line 4 × line 5) $__________

Additional costs

Cleanup ______________________________ $__________

Replacement tree installation ______________________________ $__________

Aftercare ______________________________ $__________

12. Total additional costs $__________
13. Total reproduction cost (line 11 + line 12) $__________
14. Rounded $__________

* Diameter and cross-sectional area may be replaced with plant area, volume, or height as appropriate.
[‡] Apply depreciation if it is appropriate for the assignment.

Council of Tree & Landscape Appraisers (CTLA). 2019. *Guide for Plant Appraisal, 10th Edition*. International Society of Arboriculture, Atlanta, GA.

Functional Replacement Method
Trunk Formula Technique

Client name __________ Date ______ Case # ______

Phone __________ E-mail __________

Address __________

Subject tree

Species __________

1. Trunk diameter* (D) ______ in. @ ______
2. Condition rating ______ %

   Health __________ ______

   Structure __________ ______

   Form __________ ______
3. Functional limitations __________ ______ %
4. External limitations __________ ______ %

Functional replacement tree

Utility or benefit to be replaced __________

Replacement plan __________

5. Trunk diameter* (D) ______ in. @ ______
6. Cross-sectional area (line 5)^2 × 0.7854 = ______ in^2

Replacement nursery tree

7. Trunk diameter* (D) ______ in. @ ______
8. Cross-sectional area (line 7)^2 × 0.7854 = ______ in^2
9. Nursery tree cost Source: __________ $______

Calculations

10. Unit nursery tree cost (line 9 ÷ line 8 or from RPAC) $______ /in^2
11. Basic functional replacement cost (line 6 × line 10) $______
12. Depreciated basic cost† (line 11 × line 2 × line 3 × line 4) $______

Additional costs

Cleanup __________ $______

Nursery tree installation __________ $______

Aftercare __________ $______

Hardscape __________ $______

Other __________ $______

13. Total additional costs† (sum additional costs) $______

Total functional replacement cost (line 11 or 12 + line 13) $______

Rounded $______

* Diameter and cross-sectional area may be replaced with plant area, volume, or height as appropriate.

† Apply depreciation and add additional costs if appropriate for the assignment.

Council of Tree & Landscape Appraisers (CTLA). 2019. *Guide for Plant Appraisal, 10th Edition*. International Society of Arboriculture, Atlanta, GA.

Reproduction Method
Cost Compounding Technique

Client name ______________________________ Date ________ Case # __________

Phone ______________________________ E-mail ______________________________

Address ____

Subject tree

Species ______________________________

1. Trunk diameter* (D) ______________ @ ____________
2. Condition rating ________ %

   Health ______________________________

   Structure ______________________________

   Form ______________________________
3. Functional limitations ______________________________ ________ %
4. External limitations ______________________________ ________ %

Replacement tree

Species ______________________________

5. Trunk diameter* (D) ______________ @ ____________
6. Replacement tree and installation cost (Source: ________________) $________
7. Site preparation (if any) and present value of aftercare (if any) ____________ $________
8. Total replacement tree cost $________

Calculations

9. Years to parity (appraiser's judgment)** ________ yrs
10. Interest rate Source: ______________________________ ________ %
11. Basic compounded cost (line 8 × $[1+ \text{line } 10]^{\text{line } 9}$) $________
12. Depreciated compounded cost‡ (line 11 × line 2 × line 3 × line 4) $________
13. Additional clean-up cost $________
14. Total (line 12 + line 13) $________
15. Rounded $________

* Diameter and cross-sectional area may be replaced with plant area, volume, or height as appropriate.

** The age and cross-sectional area of the subject tree are not necessarily relevant. Its size (diameter, volume, and/or height) is relevant. Years to parity should reflect the appraiser's best estimate at the time for a healthy specimen to reach a size where it provides equal utility or benefits.

‡ Apply depreciation if it is appropriate for the assignment.

Functional Replacement Method
Cost Compounding Technique

Client name ______________________ Date ________ Case # ________

Phone ______________________ E-mail ______________________

Address ______________________

Subject tree

Species ______________________

1. Trunk diameter* (D) ____________ @ ____________
2. Condition rating ________ %

   Health ______________________

   Structure ______________________

   Form ______________________
3. Functional limitations ______________________ ________ %
4. External limitations ______________________ ________ %

Functional replacement tree

Utility or benefit to be replaced ______________________

Replacement plan ______________________

Species ______________________

5. Diameter (D) ________ @ ________
6. Replacement tree and installation cost (Source: ______________________) $________
7. Site preparation (if any) and present value of aftercare (if any) ____________ $________
8. Total replacement tree cost $________

Computations

9. Years to parity (appraiser's judgment)** ________ yrs
10. Interest rate (Source: ______________________) ________ %
11. Basic compounded cost (line 8 × $[1+ \text{line } 10]^{\text{line } 9}$) $________
12. Depreciated compounded cost† (line 11 × line 2 × line 3 × line 4) $________
13. Additional clean-up cost $________
14. Total (line 12 + line 13) $________
15. Rounded $________

* Diameter and cross-sectional area may be replaced with plant area, volume, or height as appropriate.

** The age and cross-sectional area of the subject tree are not necessarily relevant. Its size (diameter, volume, and/or height) is relevant. Years to parity should reflect the appraiser's best estimate of the time for a healthy specimen to reach a size where it provides equal utility or benefits.

† Apply depreciation if it is appropriate for the assignment.

Council of Tree & Landscape Appraisers (CTLA). 2019. *Guide for Plant Appraisal, 10th Edition.* International Society of Arboriculture, Atlanta, GA.

CHAPTER 6

The Income Approach

CHAPTER OUTLINE

Overview

The income approach (also called the income capitalization approach) is used to appraise income-producing plants or property. Appraisal problems that call for the income approach fall into two categories: (1) income-producing properties such as nurseries, Christmas tree farms, orchards, vineyards, timber, and forestland; and (2) value associated with ecological benefits.

The basis of the income approach is the principle of anticipation, the idea that value is derived from the expectation of future benefits (income). Because the income approach looks to the future, the time value of money is a key aspect. Money available today is worth more than the same amount made available in the future due to its potential earning capacity. It is the same reason why interest is paid in savings or other investments.

This chapter presents an overview of the income approach and the direct capitalization and yield capitalization methods. Both methods estimate the present value of future income. Within these methods are various techniques (Figure 6.1).

Direct Capitalization

Direct capitalization estimates the current market value of projected income using the formula

$$V = I \div CR$$

where V = current market value, I = expected year-one income minus operating expenses (net operating income), and CR = **capitalization rate** (or **cap rate**), or the current or desired rate of return on the investment.

Net operating income (NOI) is the expected annual income minus anticipated fixed and variable expenses. Income may come from the sale of products (trees, containerized plants, or benefits) or from rent. Expenses

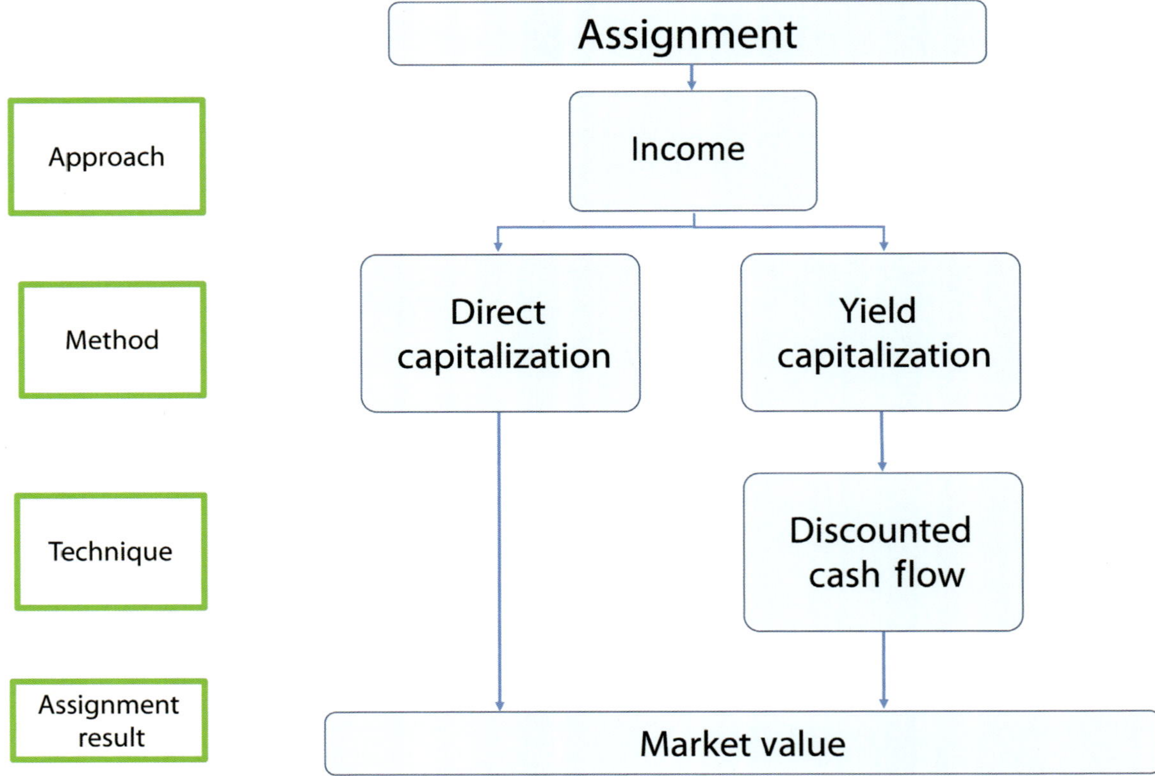

Figure 6.1 Flow chart of the income approach.

can be management fees, property taxes, insurance, maintenance, planting, utilities, or others necessary for the business. Income taxes and mortgage interest are usually not considered in direct capitalization.

Appraisers extract capitalization rates from market transactions by dividing NOI by property sale price. The resulting capitalization rate reflects the investor's expectations regarding future net income and the annual compound rate at which the property is expected to rise or fall.

Selecting the appropriate cap rate is a critical decision, as the outcome is very sensitive to the rate used. A cap rate may be determined for the business or it may be assigned by an investor to aid in purchasing decisions. The cap rate reflects the trade-off between risk and return. The higher the perceived risk, the greater the rate of return must be to attract capital to the investment. If a business is well established and has a solid client base, it would justify a lower cap rate more than a younger business with a few large accounts. Establishing a capitalization rate is discussed in Appendix 6.

Consider a fruit orchard that has been in business for decades and has a dependable market for the fruit. The appraiser expects the orchard to produce a year-one NOI of $23,000. The appraiser researches three local orchard sales, and the ratio of expected NOI to sale price ranges from 7.0% to 8.0% and averages 7.3%. The risk associated with the subject NOI is slightly higher than average for the three transactions, and the appraiser concludes that a rate of 7.5% is appropriate for the appraisal. Therefore, the estimated value of the orchard is $23,000 ÷ 0.075 = $306,666, which rounds to $307,000.

Where the marketplace expects material appreciation in the value of similar properties, the cap rate is lower (i.e., lower rates produce higher values). The cap rate is therefore the yield rate (required rate of return) less annual appreciation. If the yield rate is 8.0% and the expected rate of appreciation 0.5%, the cap rate is 8.0% - 0.5% = 7.5%. Conversely, if orchards are expected to depreciate by 5% annually, the cap rate is 8.0% + 0.5% = 8.5%, and dividing by the higher rate will produce a lower market value.

Discounted Cash Flow Analysis

Where projected cash flows are highly variable from year to year, and particularly where NOI is negative in some years and positive in others, **discounted cash flow (DCF) analysis** is more appropriate. It is a form of yield capitalization. It may be defined as the present value of expected cash flows, including annual or periodic NOI plus the resale (reversion) value of the property at the end of the projected holding period. Projected NOI is converted to present value by discounting annual cash flows at a market discount (yield) rate reflecting the risk of the cash flows. As with direct capitalization, the **discount rate** is derived from market transactions.

DCF analysis typically involves three basic steps: (1) estimate NOI for each year of the projected holding period, (2) derive an appropriate discount rate, and (3) discount the projected cash flows (Table 6.1). Alternatively, the appraiser can compute the present value for each year's cash flows and then sum up all the present values (see Example 4).

Table 6.1 Steps for applying discounted cash flow (DCF) analysis.

1. Define the holding period.
   - Definite
   - Perpetual or indefinite
2. Project annual revenues.
   - Gross income from services or products
3. Project annual costs.
   - Fixed costs (carrying charges, administration/overhead)
   - Variable costs (labor, commissions, management operations)
4. Derive the appropriate discount rate.
5. Convert projected net operating income (NOI) into present value:

   $$PV = FV \div (1 + r)^n$$

   where PV = present value, FV = future value, r = discount rate, and n = number of years.
6. Total the present values to estimate the discounted cash flow.

The present value of NOI for any given year can be computed using the following formula:

$$PV = FV \div (1 + r)^n$$

where PV = present value, FV = future value, r = discount rate, and n = number of years.

The estimate of present value relies on a discount rate. A discount rate is similar to the cap rate. It is the interest rate used to determine the present value of future cash flows.

The discount rate may be derived from transaction analysis, or it may be based on a common financial instrument like treasury bills or stocks that have a similar risk as the item being appraised. Like the cap rate, the discount rate is a direct expression of the time value of money: the riskier the investment, the higher the discount rate. Unlike the cap rate, the discount rate does not account for an expected change in property value over the holding period. The appreciation or depreciation in property value is expressed in the final sale price (reversion) as a part of the final year's cash flows.

For a string of projected future net cash flows (NOIs) occurring at the end of each year, the formula for present value is:

$$DCF = CF_1 \div (1 + r)^1 + CF_2 \div (1 + r)^2 + \ldots + CF_n \div (1 + r)^n$$

where CF = cash flow or net operating income, r = discount rate, and n = number of years.

Appraisers apply this technique with timber, Christmas tree farms, and other properties or businesses where cash flow is expected to vary considerably from year to year. They may also apply DCF analysis to annual cash crops from orchards and vineyards.

In cases like commercial timberlands, there may be a series of harvest cycles (rotations). If there is an uneven distribution of age classes, the net operating income will rise and fall for each harvest period. In addition to income from the harvest, there may be income from the sale of the land or other assets at the end of the period. In the classic application of DCF, the reversion value is the present value of expected income from the one-time sale of all the timber, or from the sale of the land after one or more harvests.

For example, a timber company may apply DCF analysis when entering a ten-year timber lease where the quantity of timber available for harvest and price for the timber is anticipated to change during the lease period. The appraiser should forecast net operating income for each year. These calculations need to consider a range of variable costs (management fees, sales commissions), annual carrying charges (e.g., property taxes, annual rent), overhead costs, and changing timber prices. There may be no reversion beyond harvest income in this case, as the lease simply expires.

Applying DCF to i-Tree Analysis

The annual benefits provided by urban trees are a summation of partial benefits (Peterson and Straka 2012), including but not limited to aesthetics, social and psychological well-being, wildlife habitat, public health, energy savings, reduction in atmospheric contaminants, reduction in stormwater runoff, contribution to property value, and carbon sequestration. The value of some benefits can be established by markets that trade in them. In addition to benefits, there are costs associated with managing trees in urban areas, including planting, pruning, removal, pest management, and litigation. The value of the net benefits can be viewed as a revenue stream provided by the tree. The potential revenue stream is not infinite but falls off as the tree ages and approaches the end of its life span.

i-Tree Eco (www.itreetools.org/eco/) provides estimates of the annual value of some of the benefits provided by a tree or group of trees over a future number of years. The forecast module of i-Tree Eco allows more flexibility in defining the tree's life span. From i-Tree Eco's forecast module, the net present value of anticipated benefits can be calculated using a discount rate.

The actual cash flows relating to i-Tree Eco calculations may not be directly realized. Energy savings in the form of reduced heating and/or cooling costs go directly to the homeowner. The value of atmospheric contaminant removal, CO_2 sequestration, and stormwater reduction does not. These values are provided to municipalities and society as avoided costs. For example, a tree removes atmospheric contaminants. The municipality does not have to pay to remove those contaminants or for health care for those affected by the contaminants. It is an avoided cost, associated with reduced healthcare expenditures, reduced damage to materials, reduced irritation, and increased life spans due to cleaner air.

One issue with projecting future value is the forecast period that reflects the tree's remaining life span. Estimates of life span for trees can be based on average mortality rates for a tree population (if known) or estimates of current tree age versus expected life span for a tree species. As with any DCF analysis, the result can be very sensitive to the interest (discount) rate.

STRENGTHS AND LIMITATIONS OF DISCOUNTED CASH FLOW AND i-TREE

Strengths

- Benefit values are based on regional market information with frequent updates.
- The calculations are easily performed.
- It has a strong research foundation.
- DCF is a well-accepted procedure.

Limitations

- Results are sensitive to estimates of life span and interest rates.

Applying DCF to Business Decisions

The income approach can be used to assist the owner of an income-producing property to decide if it should be held or sold. For this type of evaluation, the income approach is combined with the sales comparison approach. For example, an investor would retain an appraiser to use the income approach to estimate how much a property is worth in terms of its net present value. The appraiser would then use the sales comparison approach to test the results against the market value of the property. If the market promises to pay more for the sale of the land than it will realize in net operating income over time, they may choose to sell.

If the direct capitalization method is applied to an income-producing timberland property that has a net operating income of $35,000, and a rate of return of 10% is also applied, the resulting value would be $350,000, where $35,000 ÷ 10% = $350,000. If the sales comparison approach indicates a market value of $300,000 for the same income-producing property, the owner of the timberland property may choose to hold the property and continue to operate it.

Depreciation

In contrast to the cost approach, the income approach accounts for depreciation in the projections of revenues and operating expenses. If plants are in poor condition and this results in reduced profits, this will be expressed in lower revenues and/or higher costs. The income approach is entirely focused on the capacity of the property to produce net operating income. Everything that affects future cash flows and present value is reflected in projected net operating income and in the capitalization rate or discount rate used to convert cash flows into present value.

Examples Using the Income Approach

Example 1. Nursery Affected by Highway Relocation

A new four-lane highway will take one-third of the land area and planted material from Moon Pine Nursery. The nursery has an expected year-one income of $150,000 and operating costs of $135,000, for a projected NOI of $15,000. The appraiser hired by the state department of transportation estimates the depreciated reproduction cost of the shrubs and trees at $35,000 and the value of the bare land at $6,000, for a combined value of $41,000. Mr. Moon thinks the loss to his business is greater than the sum of the plants plus the land. He hires an appraiser for a second opinion.

Negotiations or litigation may ultimately decide the actual damages in this example, but the income approach, using the direct capitalization method, can provide an analysis of the change of value in his business. With one-third of his land and nursery income lost because of the highway expansion, the owner's argument is that his annual net income will also dwindle from $15,000 to $10,000.

Mr. Moon hires a knowledgeable appraiser to estimate the current value of the nursery. The appraiser looks at other nurseries that have recently sold in the area to determine the market value using the sales comparison approach. The appraiser develops an appraisal of $156,250 for Moon Pine Nursery. To appraise the value of the business before and after the taking, the appraiser employs the direct capitalization method. After looking at NOI, the appraiser calculates a cap rate prior to the taking:

$$\begin{aligned} \text{cap rate} &= \text{NOI} \div \text{value} \\ &= \$15{,}000 \div \$156{,}250 \\ &= 0.096 \\ &= 9.6\% \end{aligned}$$

The appraiser then estimates the value after the taking, assuming an NOI of $10,000 and a cap rate of 9.6%:

$$\begin{aligned} \text{value} &= \text{NOI} \div \text{cap rate} \\ &= \$10{,}000 \div 0.096 \\ &= \$104{,}167 \end{aligned}$$

Therefore, the loss of value is $156,250 - $104,167 = $52,083, which is greater than the $41,000 offered by the DOT.

An alternative solution would be to

1. derive a market cap rate from the comp sales;
2. divide before and after NOIs by the cap rate to compute before and after present values; and
3. subtract after value from before value.

If the cap rate derived from the comp sales is the same as the 9.6% computed above, the result will be the same.

Example 2. Apartment Value Damaged by Removal of Specimen Shade Trees

An exclusive apartment complex commands a view of an allée of stately oaks. The oaks are over 100 years old and range in dbh from 30 to 36 inches (75 to 90 cm). They provide benefits to the complex, including a pleasant view, shade, and a place for residents to picnic, stroll, and relax. The trees lie along the west property boundary. The complex has ten units with a very low vacancy rate. Individual apartments command an additional $25 per month compared to similar units in the area.

An adjoining property owner breaks ground on a new development. The construction company crosses the property line and removes all the oak trees. The owner of the adjacent property is found at fault for damaging the apartment property. Two plant appraisers are called in to investigate and estimate property damages. The plaintiff's appraiser estimates damages using the reproduction cost method at $50,000.

The defendant's appraiser uses similar cost approaches and arrives at a reproduction cost estimate of $45,000. The appraiser also applies the income approach, focusing on lost revenues. The appraiser investigates the rent history and concludes that the value of the oaks has contributed $25 per month for each unit. The total annual loss in rent is $3,000 ($25 per month at ten units for 12 months).

Recent sales of similar properties indicate that the average owner or investor holds an apartment building for ten years, and cap rates average 10%. Rent increases are keeping pace with inflation, and this is reflected in the 10%. Using this information, the appraiser estimates the value of the oaks:

$$\begin{aligned} \text{value} &= \text{NOI} \div \text{cap rate} \\ &= \$3{,}000 \div 0.10 \\ &= \$30{,}000 \end{aligned}$$

where the NOI is $3,000 and the cap rate is 10%.

The defendant's appraiser weighs all the evidence and concludes that the income estimate of $30,000 is a more reliable damage estimate. Both appraisals are presented in court. The plaintiff seeks damages of $50,000. The defendant responds with a damage estimate of $30,000.

Example 3. Fire Kills Maple Trees in Commercial Sugar Bush (adapted from Steigerwaldt and Steigerwaldt 2012)

A brush fire spreads from a neighbor's property to an adjacent stand of sugar maple (*Acer saccharum*) trees. The stand has a history of sugar production and is insured. Based on the terms of the insurance policy, the insurance company asks an appraiser to estimate the economic loss to the owner.

The appraiser finds that 86 trees had more than 75% of their trunk circumference seriously damaged and concludes that these trees will either die or become nonproductive. The appraiser's analysis produces the following information:

- Syrup producers pay $1.25 to $1.75 per tap annually for collection rights.
- The sugar lease allows a maximum of 120 taps in the 86 trees.
- The typical cap rate for maple syrup operations in this area is 4% to 8%.

The landowner asserts that he earns $28 per tree annually, based on retail maple syrup sales. The appraiser also recognizes that the trees could be used for commercial timber and develops the following information for a sales comparison market value:

- Each tree has an average volume of 100 board feet of sawtimber plus one-tenth cord of pulpwood.
- The average maple stumpage is $300 per 1,000 board feet, and the pulpwood stumpage is $20 per cord.

As summarized below, the appraiser must select between two indications of value: one based on maple syrup tap leases, and one based on timber income.

The value as tap lease can be calculated as follows:

120 taps × $1.50/tap = $180 income per year
Using a mean cap rate of 6%, the present value is $180 ÷ 0.06 = $3,000
Value per tree = $3,000 ÷ 86 = $34.88

The value as timber can be calculated thusly:

100 board feet sawtimber per tree × 86 trees = 8,600 board feet
$300/1,000 board feet × 8,600 = $2,580
$20/cord × 8.6 cords = $172
Total = $2,752
Value per tree = $2,752 ÷ 86 = $32.00

The tap-based income yields a higher estimate and thus indicates that this is the highest and best use of the property and the better indication of damages.

Example 4. Estimating the Present Value of Future Benefits Using i-Tree

The plant appraiser is contacted by a landscape architect to assess the benefits of installing an American elm tree (*Ulmus americana*) near the new residence. The appraiser uses i-Tree Design to place a 2-inch-diameter (5.08-cm) tree in the front yard. The program estimates that the tree will produce one dollar in annual benefits at planting. After 20 years, the tree is predicted to be 17.5 inches (44.5 cm) in diameter, providing $49 in annual benefits.

Suppose the tree was killed after 20 years. What would be the present value of the benefits that would accrue over the next 30 years with a discount rate of 3.5%?

$$PV = \$49 \times ([1 - 1.035^{-30}] \div 0.035)$$
$$= \$901.21$$

Example 5. Appraising the Economic Impact of Damage to a Tree

The appraiser is contacted by a home owner whose tree was damaged by a motorist who ran off the road. The tree's estimated replacement cost is $600. The tree was in good health before the accident with a useful life expectancy of 60 years. After the accident, the appraiser judged the remaining useful life expectancy to be ten years. One way to measure the economic impact of the damage is to compute the difference in present value of the expected replacement cost before and after the accident. The reasoning is that the damage has placed the homeowner in the position of having to replace the tree in 10 years instead of 50 years, and the time value of money measures the economic loss or opportunity cost of having to invest $600 sooner than was needed prior to the accident. The present value of the expected replacement cost is

$$PV = FV \div (1 + i)^n$$

where PV = present value, FV= future value, n = the number of years, and i = discount (interest) rate.

Using a home mortgage rate of 7.0%, the present value of replacing the tree in 50 years is computed as follows:

$$PV = \$600 \div (1 + 0.07)^{50}$$
$$= \$600 \div (29.457)$$
$$= \$20.37$$

Therefore, the present value of $600 to be paid in year 50 is only one twenty-ninth or 3.4% of $600.

The present value of replacing the tree in ten years is

$$PV = \$600 \div (1 + 0.07)^{10}$$
$$= 600 \div 1.967$$
$$= \$305.01$$

Therefore, the damages amount to $285 ($305 - $20).

To this point, this example assumes that the replacement cost of the tree is stable over time. However, replacement costs will increase over time. For example, if the cost to replace the tree is expected to rise by 3% per year, the future value of the replacement costs are as follows:

Future value (FV) = $PV(1 + i)^n$
Cost to replace in 50 years ($V50$) = $\$600 \times (1.03)^{50} = \$600 \times (4.384) = \$2{,}630$
Cost to replace in 10 years ($V10$) = $\$600 \times (1.03)^{10} = \$600 \times (1.344) = \$806$

Inflating costs causes the difference between these two figures to increase, so one would expect the economic damages to be greater than in the case of stable costs. The present value of replacing the tree in 50 years is:

$$V_0 = \$2{,}630 \div 1.07^{50}$$
$$= \$2{,}630 \div 29.457$$
$$= \$89.29$$

The present value of replacing the tree in 10 years is

$$V_0 = \$806 \div 1.07^{10}$$
$$= \$806 \div 1.967$$
$$= \$409.91$$

Therefore, the damages amount to $410 − $89 = $321.

CHAPTER 7

The Sales Comparison Approach

CHAPTER OUTLINE

Overview

The sales comparison approach (SCA) identifies sale transactions (sales history, comparable sales, listing prices, etc.) and analyzes those transactions to develop an opinion of market value. Tree and landscape appraisers can use information from real estate transactions, tax assessments, and third-party real estate appraisals as the basis for estimating the value of landscape items or individual trees. Since this approach uses empirical (verifiable by observation) data, it gives the SCA widespread credibility and makes it the most common approach for most types of real estate.

There are two guiding principles with the SCA: (1) the principle of substitution, i.e., prudent and knowledgeable buyers acting in their own self-interest will not pay more than the price of an equivalent substitute; and (2) the principle of contribution, i.e., the value of a particular component of a property can be measured in terms of its contribution to the value of the whole property, or the amount by which its absence would detract from the value of the whole (Chapter 2). Trees and other landscape items generally add value to developed property (Figure 7.1).

Applying the SCA is most straightforward where there are records of comparable property sales with and without the item. The appraiser then can isolate the contribution made by a single tree or other landscape feature. In practice, however, it is difficult to find comparable sales that provide clear evidence of this contributory value. Buyers and sellers of real property tend to think about the property as a whole, as opposed to considering the sum of individual parts. The situation is further complicated because buyer preferences vary greatly. The presence or absence of a tree may not affect the market value of a property to any given buyer.

These challenges usually do not apply to trees where the highest and best use is for commercial timber. There is a unique market for such trees. The appraiser identifies comparable stumpage sales or sales to a mill and applies the unit rate of the timber (Chapter 9).

Figure 7.1 Properties with trees and high-quality landscapes typically sell for more than equivalent properties without landscapes.

HIGHEST AND BEST USE (HBU)

Highest and best use should be considered a function of the appraisal problem, regardless of which approach is applied. The sales comparison approach, however, requires the appraiser to think in terms of transactions-oriented evidence, which produces the most empirical evidence of a property's HBU. Chapter 2 presents the definition and criteria for HBU.

For residential and commercial landscape applications, there are usually no records of tree or landscape value. Therefore, component analysis becomes the fundamental method in SCA. Three component analysis techniques are available to landscape appraisers: (1) **hedonic regression analysis**, (2) direct or paired comparison, and (3) extraction. In addition, some of these techniques can be paired with the cost approach to estimate market value.

Landscape appraisal assignments that typically call for the SCA include the following:

- estimating current or diminished property value for insurance, condemnation actions, litigation, and income tax, gift tax, and inheritance tax purposes;
- analyzing the contribution of plants to overall property value;
- testing the reasonableness of a cost estimate by relating it to the overall value of the real estate; and
- valuing commercial timber crop in the context of transaction analysis, timber trespass, and other situations.

This chapter focuses on the contribution of landscape items to the market value of a real estate property (Figure 7.2). For a more detailed discussion, see Steigerwaldt and Steigerwaldt (2012).

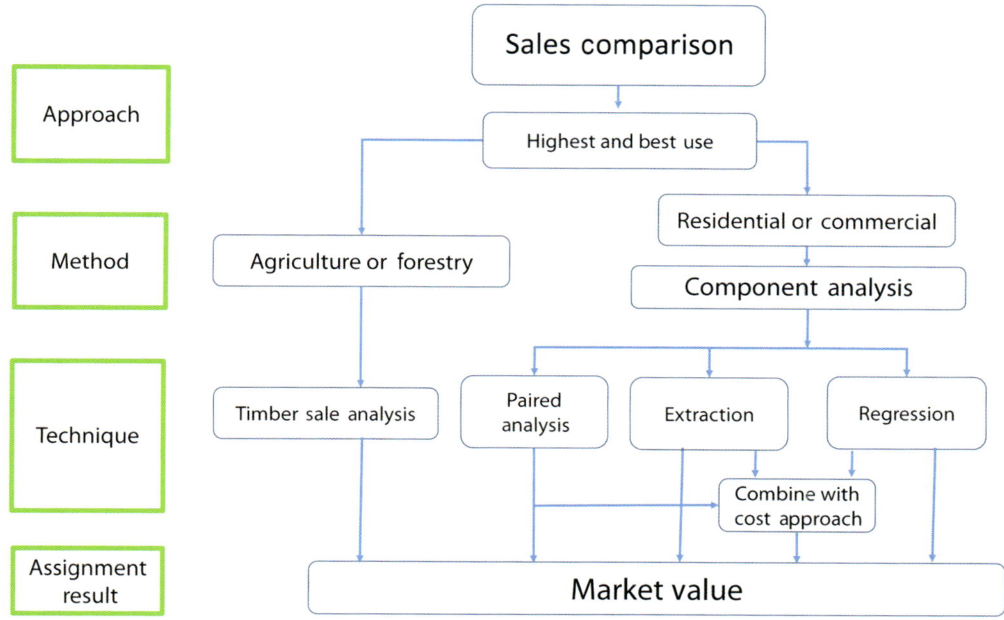

Figure 7.2 Flow chart of the sales comparison approach and its use to determine a market value.

Component Analysis Method

The SCA commonly uses the component analysis method to estimate market value. It estimates the value of a component (e.g., a tree in a landscape) by comparing similar sales of whole properties, with and without the component, and infers that the differences in sales price are due to the value of the component. This process becomes difficult when the property contains multiple features that could affect the sales price, such as lot size, house size, view, neighborhood desirability, driveway pavement, or landscape features.

1. Hedonic Regression Analysis Technique

Regression analysis computes the statistical correlation between an independent or causal variable and a dependent variable. Where more than one variable is evaluated, **multiple regression analysis** is used. Regression analysis is used to explain or predict the relationship between the independent and dependent variable(s), like the contributory value of landscaping to overall property value (Appraisal Institute 2015: p. 330). In such an analysis, independent variables may include total property value, lot value, lot size, house size, location, and landscape factors. Hedonic regression analysis evaluates the demand for the constituent components of overall property value and estimates the contributory value of each. Hedonic regression analysis identifies which characteristics are significant to buyers and how much they will pay for them. It is actually a variation of the SCA insofar as it can be used to predict value from statistically significant variables that affect value. In hedonic price analysis, regression is used to quantify buyer preferences for various characteristics of a property (Appendix 3).

Regression analysis is common in mass appraisal of residential property for tax valuation but is rarely used directly by landscape appraisers. However, many studies using hedonic regression analysis have been conducted for landscapes, trees, and parks (Table 7.1). Results of these studies can be applied to predict or isolate the influence of trees on property value. Because hedonic regression analysis tends to generalize over many properties, this technique may not apply well to the appraisal of an individual tree. Most of these studies do not distinguish among tree species, size, condition, or placement.

Table 7.1 Hedonic studies of the contribution of trees to urban residential values.

Value contribution	Factors considered	Research citation
0.75% to 2.5%	Mean canopy cover. Diminishing returns past a certain level of cover.	Netusil, N.R., S. Chattopadhyay, and K.F. Kovacs. 2010. Estimating the demand for tree canopy: A second stage Hedonic price analysis in Portland, Oregon. *Land Economics* 86(2):281–293.
2%	Single mature landscape tree (> 9 in dbh).	Dombrow, J., M. Rodriguez, and C.F. Sirmans. 2000. The Market Value of Mature Trees in Single-Family Housing Markets. *Appraisal Journal* 68, 1:39–43. Baton Rouge, LA.
2%	Street tree contribution to adjacent property value.	Wachter, S.M., and G. Wong. 2008. What is a tree worth? Green-city strategies and housing prices. *Real Estate Economics* 36(2):213–239.
2.4% 6.4%	Single street trees. Two street trees.	Donovan, G.H., and D.T. Butry. 2010. Trees in the City: Valuing Street Trees in Portland, Oregon. *Landscape and Urban Planning* 94, 1:77–83.
3.5% to 4.5%	Trees in front-yard landscape.	Anderson, L.M., and H.K. Cordell. 1988. Influence of Trees on Residential Property Values in Athens, Georgia. *Landscape and Urban Planning* 15:153–164.
	Total landscape contribution:	
4% to 7%	Difference from good to excellent.	Henry, M. S. 1994. The contribution of landscaping to the price of single family houses: A study of home sales in Greenville, South Carolina. *Journal of Environmental Horticulture* 12(2):65–70.
8% to 15%	Difference from poor to excellent.	Henry. 1999. Landscape Quality and the Price of Single Family Houses: Further Evidence from Home Sales in Greenville, South Carolina. *Journal of Environmental Horticulture* 17(1):25–30.
5% to 15% Mean 7%	Additional value from trees to the price of a lot.	Payne, B.R. and S. Strom. 1975. The contribution of trees to the appraised value of unimproved residential land. *Valuation* 22(2):36–45. Massachusetts. Payne, B.R. 1973. The twenty-nine tree home improvement plan. *Natural History* 82(9):74–75.
< 6%	Good tree cover.	Morales, D.J. 1980. The contribution of trees to residential property value. *Journal of Arboriculture* 6(11):305–308.
Mean 7%	Home sales on wooded lots vs. non-treed lots.	Seila, A.F., and L.M. Anderson. 1982. Estimating cost of tree preservation on residential lots. *Journal of Arboriculture* 8:182–185. Seila, A.F., and L.M. Anderson. 1984. Estimating tree preservation cost on urban residential lots in metropolitan Atlanta, Macon, GA. GA For Comm. For. Res Paper No. 48 6pp.
10% to 15%	Mature trees in high-income neighborhoods.	Theriault, M., Y. Kestens, and F. Des Rosiers. 2002. The Impact of Mature Trees on House Values and on Residential Location Choices in Quebec City. In: Rizzoli, A.E., and A.J. Jakeman (eds.), Integrated Assessment and Decision Support, Proceedings of the First Biennial Meeting of the International Environmental Modeling and Software Society, Volume 2:478–483.

These studies show that a single tree rarely contributes more than 2% to the value of developed residential property, and a high-quality landscape, in total, may contribute 4% to 10%.

The results of hedonic regression studies may be applied in the reconciliation phase of the appraisal to verify estimates derived from other approaches, methods, or techniques. This is done by multiplying the property value by the most appropriate percentage contribution (Table 7.1).

SOURCES FOR MARKET VALUE OF PROPERTY

Sources for market value of property include:

- municipal tax records, which separate land and building components;
- online automated valuation applications (e.g., Zillow.com, Realtor.com, Trulia.com);
- local brokers and real estate appraisers; and
- information obtained from the property owner.

The context of the assignment will determine what degree of diligence is required.

2. Paired Sales Analysis Technique

Paired sales analysis is common in real estate appraisal. In the context of plant appraisal, comparisons may involve undeveloped properties with and without trees or developed properties with varying amounts or quality of landscaping. The appraiser identifies a series of comparable sales that are alike except in one characteristic. This type of analysis is particularly useful if there are a number of nearby houses all of similar size and on similarly sized lots so that the main difference among properties are the trees or landscaping. The appraised value of each property with and without the landscape feature can be compared.

For example, a well-treed residential lot with no house or other improvements can be compared to a nearby lot with no trees. Here the difference in the unit price (dollars per acre, per square foot, or per square meter) of the two lots is the contributory value of the trees. The primary weakness of this method is the limitation in sample size (in this case, a single pair of sales) and the assumption that the difference in price reflects the buyers' preference for trees versus some other factor. It is often impossible to verify the impact of a landscape component on the price paid without interviewing the buyer to learn whether trees, or the lack of them, influenced the price.

When appraising a single shade tree that dominates the landscape, the appraiser may be able to find two residential property sales, alike in almost every aspect, but differing in the presence of one large shade tree. For example, if the value of the property with the tree is $400,000 and the value of the property without the tree is $395,000, the appraiser might deduce that the contributory value of the tree is $5,000. Where the building improvements on the property differ, the appraiser can deduct the contributory value of the building improvements by consulting tax records or some other source, leaving a residual lot value and thus simplifying the paired analysis.

Landscape appraisers can conduct simple direct comparisons to estimate tree or landscape contribution to property in their area on their own. However, it may be expedient to collaborate with an experienced local broker or a real estate appraiser when doing paired sales analysis. Situations where a property owner seeks a casualty loss claim for insurance or federal income tax purposes provide excellent opportunities for skilled plant appraisers to collaborate with real estate appraisers.

3. Extraction Technique

The extraction technique subtracts land value and all known improvements (house, outbuilding, utilities, driveway, etc.) from the overall property value, leaving the market value of the landscape as the remainder. This can be further refined by subtracting the market value (not cost) of any landscape components where the value is known.

The sales comparison approach can also be combined with the cost approach to provide a tree value. The important concept is that the contributory value of the landscape components are based on the overall property value as estimated by transactions analysis (see examples in chapters 5 and 8). Estimates developed in this fashion may be somewhat subjective but should be reasonable as measured in terms of contribution to the whole.

STRENGTHS AND LIMITATIONS OF SALES COMPARISON APPROACH

Strengths

- It ties tree and landscape value to overall property value.
- It provides a logical, defined process where the sum of the parts equals the whole.
- It uses observations of actual sales to provide a foundation for estimating the market value.
- It accounts for some forms of depreciation.
- It can be used with the cost approach to develop market value or add credibility to a cost estimate.

Limitations

- It can be difficult to extract information about a single tree or landscape feature.
- Results of hedonic regression studies are based on data outside the subject neighborhood and may not account for unique features of the subject property and its particular landscape items.
- Allocating total landscape value among specific landscape features can be subjective.
- It requires using data from real estate appraisers, brokers, assessors, tax records, or online sources.
- It assumes that landscape items contribute positive value to a property, when in fact they may not (e.g., consumer preference, superadequacy, fire hazard).

Summary

The sales comparison approach relies on the actions of buyers and sellers to estimate value. Information about transactions is available from a number of sources. Analysis of transactions allows the plant appraiser to allocate value among the land and improvements, including trees and landscape. In general, trees and landscapes have a net positive effect on real estate value.

Examples Using the Sales Comparison Approach

Example 1. Applying Results of Hedonic Regression

A plant appraiser is asked to estimate the market value of a single tree that is 12 inches (30.48 cm) in dbh and is located in the front yard of a single-family residential property. From tax records and an online real estate search, the appraiser finds that the property is appraised at $375,000. The appraiser checks the literature (Table 7.1) for studies analyzing contributory tree values in similar situations. Dombrow et al. (2000) is most similar to the situation. Results indicate that a single tree greater than 9 inches (22.86 cm) in dbh would typically contribute 2% to the market value.

Tree contribution value = property value × tree contribution percentage
= $375,000 × 0.02
= $7,500

If instead the property had five healthy trees in the front yard, the typical contributory value of all five trees based on Henry (1994) could be estimated to be between $15,000 (0.04 × $375,000) and $26,250 (0.07 × $375,000) or $3,000 to $5,250 per tree. The differences between estimates generated from the single tree analysis and the multiple tree analysis reflects the law of diminishing returns (Chapter 2) in terms of the contribution of each additional tree.

Example 2. Paired Sales Analysis: Fire Loss

A fire destroys all the trees on several undeveloped (forested) lots in a new subdivision. The owner of the lots hires an appraiser to estimate the decrease in market value in order to make an insurance or tax claim. The appraiser inspects the site and determines that the highest and best use of the forested land is residential development. The appraiser reviews online records of lot sales in nearby developments and finds ten properties that have sold recently. The appraiser verifies that the lots were wooded at the time they were sold by reviewing photographic records (Google Earth) and discovers that there were three levels of vegetation on the sold lots: (1)

Table 7.2 Market value allocation to six trees from percent of cost.

	MARKET VALUE BY PLACEMENT ALONE			
Placement	**Ratio**	**Contributory value per yard**	**Number of trees**	**Value per tree**
Front	50%	$3,500	2	$1,750
Side	30%	$2,100	2	$1,050
Rear	20%	$1,400	2	$700
Total	**100%**	**$7,000**	**6**	**$1,167**

	MARKET VALUE BY PERCENT OF COST			
Tree number	**Placement**	**Depreciated cost**	**Percent**	**Contributory value**
1	Front	$9,600	28.7%	$2,009
2	Front	$6,600	19.7%	$1,381
3	Side	$5,500	16.4%	$1,151
4	Side	$4,950	14.8%	$1,036
5	Rear	$3,850	11.5%	$806
6	Rear	$2,950	8.8%	$617
Total		**$33,450**	**100%**	**$7,000**

no trees, (2) forested like the lot in question, and (3) few trees (could not be considered forested). An analysis of sales prices for lots with trees (an average of $75,000) and without trees (an average of $60,000) indicates there was an average price difference of $15,000 per lot ($75,000 - $60,000 = $15,000).

From this data, the appraiser estimates that forested lots are selling for $15,000 more than bare lots, a 20% difference in value. This is the appraiser's per-lot estimate of value lost to the fire.

Example 3. Paired Sales Analysis: Natural Gas Leak

A natural gas leak kills all of the landscape plants on Mr. Brewer's property, which is located in a modest residential neighborhood of a major Sunbelt city. His lawyer suggests that he hire a landscape appraiser to estimate the loss in market value so that they can document a claim against the gas utility. Mr. Brewer hires an appraiser, who compares two nearby and very similar quarter-acre properties that sold before the gas leak occurred.

Property A is a three-bedroom ranch with a well-maintained landscape and sold for $200,000. Property tax assessment records indicate that the building was worth about $125,000 and the land had a value of $75,000. Property B is a two-bedroom ranch on a similarly sized lot with virtually no shrubs, trees, or evidence of professional landscaping. It sold for $165,000. Property tax assessment records indicate that the building on property B was worth $100,000 and the land had a value of $65,000. The paired sales indicate that landscaping contributed about $10,000 ($75,000 - $65,000 land value) to sale A, or roughly 13% of its land value. Mr. Brewer's two-bedroom house on the same-sized property has a published tax value of $180,000, with $110,000 attributed to the house and $70,000 to the land. The appraiser multiplies the land value by the paired sales comparison data to estimate the contributory value of the landscape plants as follows: $70,000 × 0.13 = $9,100.

Example 4. Extraction Technique

Six large shade trees occupy a residential property. The total value of the property is estimated by a real estate appraiser to be $200,000, including $15,000 for the land and $175,000 for buildings and other non-landscape improvements, leaving a residual value of $10,000 for the landscaping. The plant appraiser uses the cost approach to conclude a value of $3,000 for hardscape, lawn, and shrubs. This infers a value of $7,000 for the six trees. The final step is to allocate the $7,000 among the trees:

Total property value (from real estate appraiser)	$200,000
Improvements (buildings, utilities, driveway, etc.)	−$175,000
Land (from real estate appraiser)	−$15,000
Market value attributed to all landscaping	$10,000
Contribution of lawn, shrubs (from landscape contractor)	$3,000
Remainder attributed to trees	$7,000
Per-tree allocation ($7000 ÷ 6)	**$1,167**

This computation assumes that each tree contributed equally to property value. If the six trees differed in condition, functional limitations, or external limitations, then the plant appraiser could assign a contribution value for each tree. If two trees were located in the rear, side, and front yards, the plant appraiser might conclude contributory values of 20%, 30%, and 50%, respectively (Table 7.2, upper half).

Alternatively, the plant appraiser may elect to apply reproduction or functional replacement cost to develop more refined estimates of contributory value for the six trees. The plant appraiser estimates depreciated cost for each tree (including a factor for placement), and then computes market value by multiplying each cost estimate by its relative percentage of $7,000, the total value for all six trees (Table 7.2, lower half).

This sort of analysis is particularly useful for damage estimates where only a portion of the trees have been affected. For example, if only the trees in the front yard were damaged, the loss in market value would be either $3,500 (based solely on placement) or $3,390 (based on percent of cost).

CHAPTER 8

Reconciliation, Reasonableness, and Reporting

CHAPTER OUTLINE

Overview

This chapter discusses three aspects of the appraisal process: reconciliation, creating a reasonable and credible result, and reporting. They are considered after approaches, methods, and techniques have been applied to the appraisal problem. In reconciliation, the appraiser considers whether one result is more appropriate than another or if several results better address the appraisal problem. Appraisers strive for results that are reasonable and credible (see Chapter 3). Finally, reports, whether written or oral, should include specific items. These topics are discussed in the context of an example that employs a number of approaches, methods, and techniques.

Reconciliation

In its common usage, reconciliation is defined as bringing together people or ideas to produce agreement, consensus, or compromise. In appraisal work, reconciliation brings together and resolves disparate indications of value or cost into a conclusion. As noted in Chapter 3, the *Uniform Standards of Professional Appraisal Practice* (USPAP) require the appraiser to reconcile every appraisal, even when only one of the three valuation approaches is employed (The Appraisal Foundation 2016). The context of the appraisal assignment and the judgment of the appraiser will determine whether assignment results should be reconciled or simply reported.

The reconciliation process often includes the following steps:

1. Review the valuation approaches used and the reasons for using them.
2. Summarize the cost or value conclusions for each of the approaches used.
3. Explain the strengths and limitations of each approach used.
4. Describe the rationale for reconciling the approaches into a final conclusion of cost or value.

In step four, the appraiser may choose to select one approach to the exclusion of all others or to develop a weighted conclusion. Doing so involves the following criteria:

1. the appropriateness of the approaches, methods, or techniques;
2. the accuracy of the data and calculations; and
3. the quantity, reliability, or sufficiency of evidence presented relative to the specific appraisal problem.

Appropriateness

The appraiser evaluates the appropriateness of each approach, method, or technique by assessing (1) its relevance to the appraisal problem and (2) its applicability to the type of property being appraised. For example, an assignment requires estimating value loss due to a fire destroying some shade trees on an apartment complex. The client plans to use the appraisal to substantiate a casualty loss claim for federal income tax purposes. The appraiser has applied the cost, sales comparison, and income approaches. The Internal Revenue Service (IRS) requires **diminution in market value** as the measure of loss. Because this is an income-based property, the cost approach will probably be less relevant than the income or sales comparison approaches.

Where a large tree has been destroyed by trespass, a functional replacement cost estimate may be more reasonable than a reproduction cost estimate, particularly if the benefits can be restored at a much lower cost by planting several commonly available specimens. In this case, the appraiser has applied the principle of substitution.

Accuracy

The appraiser evaluates accuracy based on the correctness of information employed and the reliability of the analysis used. For example, a plant appraiser applying the sales comparison approach to extract the contributory value of landscaping may need to consult with local real estate professionals. Relying on interviews with one or two would have limited precision and reliability. In contrast, using data from five comparable property sales and then verifying the information by consulting a real estate appraiser and/or building contractor may produce a credible estimate.

When estimating the cost to reproduce a large tree, an appraiser can compare cost estimates from contractors who move large trees to an extrapolation method such as the trunk formula technique (TFT). Both techniques are subject to large variation. There may be substantial variation from contractor to contractor about the cost to transplant a large tree.

Applying the reproduction cost method to large trees where market value is sought may produce an estimate that does not represent contributory market value. However, a process that takes the result of this analysis and essentially calibrates it to overall property value can produce a more credible conclusion.

Quantity, Reliability, and Sufficiency of Evidence

Other factors being equal, the more supporting evidence, the more reliable the result is. The purpose and use of the appraisal also influences the weight afforded to each of the approaches.

Where the appraisal results in both cost and value estimates, it is generally not appropriate to attempt to reconcile costs with values, though it may be appropriate to compare and contrast them. It may also be appropriate for the appraiser to comment upon the reasonableness or applicability of cost and value estimates, based on the context of the assignment, the reliability or relevance of the analyses, and other factors.

It is often the case in plant appraisal that only the cost approach is used. The appraiser may choose to apply multiple methods and techniques within the cost approach and either reconcile them into a final result or simply report the different results.

When market value is being estimated, the appraiser should consider what a buyer and seller would likely agree on as a final selling price, and weigh employed approaches, methods, and techniques accordingly. When a cost is being estimated, the appraiser should consider what a consumer might actually spend to replace a landscape item and weight the approaches, methods, and techniques used accordingly. When income is considered, the appraiser should establish a reasonable expectation of future benefits.

Summary Example Using All Approaches: Illegal Removal of a City-Owned Street Tree

A home owner had a city-owned street tree removed. The urban forester plans to fine the property owner for the illegal removal. The city tree ordinance cites the *Guide for Plant Appraisal* as the source to establish the amount of the fine. A plant appraiser is asked to estimate the value of the tree using the methods in the *Guide*.

The tree was a 15-inch-diameter (38.1-cm) red oak (*Quercus rubra*) located in a 5-foot-wide (1.52-m) planting strip between curb and sidewalk. The adjacent sidewalk was displaced 0.5 inches (1.27 cm) on the tree side. The tree was in good condition and had been pruned the year before its removal. Red oaks of similar age and size were present on both sides of the street.

The *Guide* offers at least four ways to address the assignment:

1. Cost approach, reproduction method, trunk formula technique
2. Cost approach, functional replacement method, cost compounding technique
3. Sales comparison approach, component analysis
4. Income approach, present value of future benefits

Cost Approach, Reproduction Method, Trunk Formula Technique

The Regional Plant Appraisal Committee (RPAC) in the area recently updated costs associated with growing and installing nursery trees. For hardwoods such as red oak, the largest commonly available nursery-grown tree has a 4-inch (10.16-cm) caliper (trunk cross-sectional area of 13 square inches or 33.02 square centimeters). The cost to purchase a 4-inch-caliper tree is $400; to install it, $600.

Inputs	
Diameter of subject	15 in
Cross-sectional area, subject tree	176.7 in^2
Diameter of nursery tree	4 in
Nursery purchase price	$400
Installation cost	$600
Cross-sectional area, nursery tree	12.57 in^2
Unit cost (purchase price/in^2)	$31.83

Reproduction Cost: Trunk Formula Technique	
Basic cost	$5,625
Condition	80%
Functional limitations	80%
External limitations	100%
Depreciated cost	$3,600
Installation cost	$600
Total	**$4,200**

Cost Approach, Functional Replacement Method, Cost Compounding Technique

A 4-inch-caliper (10.16-cm) tree could be installed in the 5-foot-wide (1.52-m) planting space. Such a tree would be approximately 11 feet (3.35 m) tall at planting. The appraiser estimates that it would take 25 years for a newly planted tree to duplicate the overall benefits of the removed tree. The cost to purchase such a 4-inch tree is $400; to install it, $600. The appraiser selects an interest rate of prime (3.5%) plus 2%, which is 5.5%.

Replacement Cost: Cost Compounding Technique	
Purchase + installation	$1,000
Interest rate	5.50%
Years to parity (equiv. benefits)	25
Compounded cost	$3,813
Condition	80%
Functional limitations	80%
External limitations	100%
Depreciated cost	**$2,441**

Sales Comparison Approach, Component Analysis

The property was purchased by the owner six months ago for $325,000. Local real estate brokers report that market values have increased an average of 1% since that time, making the current market value approximately $328,250. The appraiser recently read a journal article indicating that a reasonably healthy, mature street tree in a residential area contributes an average of 2.5% to market value. The red oak was the only street tree in front of the subject property. The subject property and nearby homes were landscaped in a similar manner: street trees and a front lawn with small trees and shrubs. At 15 inches (38.1 cm) in diameter, the red oak was just reaching maturity.

Sales Comparison Approach	
Market value a month ago	$325,000
Increase in one month	1%
Market value now	$328,250
% contribution to market value	2.50%
$ contribution to market value	**$8,206**

Income Approach, Present Value of Future Benefits

The 15-inch tree could be expected to provide the same function and utility for another 40 years. A recent i-Tree Eco analysis of the city's urban forest estimated the value of ecosystem services (benefits) provided by street trees as $92 per year.

Income Approach	
Annual benefits	$92
Years of lost benefits	40
Interest rate	5.50%
Present value of lost benefits	**$1,476**

Four different estimates have been produced, each with its own strengths and weaknesses, based on assumptions and the availability of data:

Process	**Result**
Cost approach, reproduction method, trunk formula technique	$4,200
Cost approach, replacement method, cost compounding technique	$2,441
Sales approach, component analysis	$8,206
Income approach, present value of future benefits	$1,476

It would be reasonable for the appraiser to reconcile the results as part of their analysis. After considering the merits of each approach, the appraiser evaluates which, if any, of the results appears strongest. As noted previously, the appraiser has several options:

1. Select one result on the basis of professional judgment. If market value is of primary importance, the $8,206 estimate clearly reflects consumer behavior and can be substantiated by empirical transactions.
2. Weigh each result based on its merits. The first two (cost) estimates are more consistent with one another than the last two (value) estimates. A replacement cost may warrant more weight because it reflects the principle of substitution. The income approach likely warrants the least weight because it is based more on theoretical regional models than on actual consumer behavior.

Process	Result	Weight
Cost - reproduction	$4,200	30%
Cost - replacement	$2,441	40%
Sales	$8,206	20%
Income	$1,476	10%
Weighted average	**$4,025**	100%

3. Compare the cost results to the value results.

Cost	average $3,320	(range $2,441 to $4,200)
Value	average $4,841	(range $1,476 to $8,206)

Finally, the appraiser might simply report the above statistics and let the client advocate for a particular conclusion. Depending on the circumstances, there may be no right or wrong answer.

The Reasonable and Credible Appraisal

The ninth edition (Council of Tree and Landscape Appraisers 2000: p.99) introduced the topic of reasonableness as follows:

> *The* Guide *advocates that methods be used in a reasonable manner. It gives the plant appraiser information about balancing plant appraisals against actual property values in the marketplace. Tests of reasonableness may be used to compare appraised plant values to total property value.*

This section addresses aspects of reasonableness and credibility. Reasonable can be viewed as logical, empirical, objective, and sensible. Credible is believable, trustworthy, and/or supported by relevant data and sound analyses.

Plant appraisers should strive to produce reasonable and credible work products. They should conduct their work in the absence of bias or advocacy and meet high professional standards. They should take necessary steps to ensure that their work is credible in the eyes of their clients, their peers, the courts, and the general public. The appraiser who walks the narrow path of objectivity and competence will produce reasonable and credible assignment results and will communicate the results accurately and without ambiguity.

The appraisal problem and the context of the assignment should determine whether a particular appraisal methodology or conclusion is reasonable (see Chapter 3).

The issue of reasonableness often arises where concepts of substitution and market value are relevant to the appraisal assignment. As described in Chapter 2, definitions and concepts that are critical to the issue of reasonableness include the principles of substitution, contribution, balance, and consistent use; highest and best use (HBU); and the law of diminishing returns. When estimating the market value of any item of real property or personal property, the appraiser should be prepared to answer one of the following two questions:

- What price would the item command if sold on the open market?
- How much less would the property be worth in the absence of the item (or after it was damaged)?

The challenge with estimating the market value of landscape plants is that, once they are installed, there is normally no market for their sale apart from the land on which they sit.

Plant appraisers routinely opine on the value of vegetation-related personal or public benefits, such as recreation, shade, aesthetics, wildlife habitat, clean water, carbon sequestration, and soil stabilization. Several of these benefits, but not all, can be estimated using i-Tree Eco software. Other benefits are intangible and more challenging to quantify. However important the benefits of plants may be to a particular owner or to the public, there is no single best process for estimating their value that applies in all cases.

Influences on Reasonableness and Credibility

The credibility of the plant appraisal profession depends on appraisers remaining independent and objective throughout the appraisal process and conducting work to acceptable professional practices. This is the ethical side of reasonableness. Conducting work to acceptable professional standards is the technical side of reasonableness. If the appraiser is ethical and competent, and applies sound judgment, then the resulting appraisal conclusions are likely to be reasonable. A breakdown of ethical standards, technical competence, or judgment can compromise the quality or reasonableness of the appraisal.

Bias can be introduced in many ways. The client can provide the appraiser selective information about the property, the marketplace, or the case itself. The client can pressure the appraiser. Errors of both omission and commission may arise. A series of small errors that operate in a single direction can have a large effect on the appraiser's conclusions.

The American Society of Consulting Arborists (ASCA) and the Appraisal Institute both have codes of **ethics** prohibiting the appraiser from being an advocate. When individuals act as an advocate, they must not represent themselves as appraisers. USPAP does not allow a licensed real estate appraiser to be an advocate for any party to the assignment. While this advice was specifically written for real estate appraisers, it is sound, and the CTLA promotes a similar standard for plant appraisers.

The client may try to influence the appraiser's work. Where conflicts over the scope, method, or result arise, the appraiser should first attempt to reconcile the difference with the client. If the client refuses to allow the appraiser to be independent and objective, then the appraiser should refuse the assignment (or terminate the appraiser-client relationship).

Attorneys are another potential influence. The American judicial system is adversarial, and attorneys are obliged to advocate for their clients. The appraiser, however, is obliged to be independent and objective in all aspects of the valuation assignment. Appraisers may advocate only for their opinion of cost or value. Appraisers should be prepared to establish their independence if they sense that the attorney is pushing them to consider assumptions, data, methods, or analyses that favor the cause of the attorney's client. Appraisers owe the duty of independence and credibility to themselves, to users of their work product, and to the appraisal profession.

Other sources that may influence an appraiser's work include the following:

- professional association codes of ethics and/or appraisal standards
- local, state, and federal statutes and/or regulations
- courts of law
- the appraiser's employer

Who Determines What Is Reasonable?

Appraisers should have a good sense of what is reasonable and what is not. If the appraiser is not confident about his or her conclusion, then this may signal that missteps have occurred somewhere in the appraisal process. Lack of confidence can also stem from a lack of relevant data and other factors. Confidence by itself, however, does not ensure credible results.

The Attorney. In matters of litigation, a skilled attorney and plant appraiser will work together to determine which methodology is most appropriate given both the legal issues and how they might affect the methodology. Attorneys may not have the background and experience with plant appraisal to know what is appropriate. In such situations, the plant appraiser has the opportunity to educate the uninformed attorney as to what is right and what cannot be justified.

A good attorney will also be aware of legal precedent and relevant statutes and regulations, and they may direct the appraiser to use certain assumptions, methods, or analyses. The appraiser should, however, guard against blindly accepting the directives of an attorney or potentially inappropriate or irrelevant court decisions. Should the appraiser disagree with the valuation methods suggested by case law or the attorney, it may be appropriate to suggest alternative valuation methods, point the attorney to authoritative literature, or even withdraw from the assignment. Once again, the appraiser should keep in mind that the attorney is obliged to advocate for his or her client, but the appraiser is allowed to advocate only for his or her own opinions.

There are a number of resources that plant appraisers may review to complement their experience. Bonapart (2014), Stamen (1997), and Merullo and Valentine (1992) provide a general introduction to trees and the law. Bloch (2000) describes a number of legal cases involving trees in relation to issues such as boundary disputes, trespass, and personal injury.

The Courts. Court cases provide legal precedent and may offer insight into appraisal methods and results. The CTLA cautions, however, against relying too heavily on case results. Court decisions are often based on factors quite distinct from appraisal, particularly where punitive damages are awarded or where appraisal analysis advanced by experts is poor in quality or otherwise unprofessional.

Damage to property is a common area of contention about reasonableness. The litigants or parties are likely to disagree about the most appropriate way to make the injured party "whole" as well as the cost to achieve that goal. Courts have recognized various methods for measuring damages, including timber stumpage value, loss in overall property value, restoration costs, and extrapolated costs.

To demonstrate the dilemma faced by the courts, consider the following excerpt from the State Bar of Wisconsin:

> *It is difficult to value the amount of damages sustained by the owner of shade trees or other ornamental trees and shrubbery because such trees are not grown for economic reasons, but rather for aesthetic reasons. In most cases, such trees and shrubbery have no significant fair market value after they have been cut [unlike timber trees].*
>
> *A person who sustains a loss of shade trees and ornamental shrubs is not limited to a recovery of the diminished fair market value of his or her property. Even though there may be no probable loss of market value of the property, the court has recognized that such trees and shrubs may have a real value to the owner for certain purposes, and in such instances, cost of replacement is the proper measure of damages. But if brush or wild trees are damaged, the court is more inclined to use the diminished value rule. When the age and sizes of the shade trees and shrubbery lost or other circumstances are such that it is practical to restore the premises, the owner is entitled to recover in damages the amount reasonably necessary to do so. When this cannot be done (e.g., in cases involving mature shade trees and ornamental shrubs), the owner is entitled to recover the value that the things destroyed contributed to the real property, given the particular purpose of the owner's occupancy. Thus, it is necessary to evaluate the extent to which this destruction has interfered with the use of the land and with the owner's comfort and convenience.* (Ware 2005)

Legal Considerations. Statutes, case law, ordinances, and regulations differ widely across state and local jurisdictions. They will, in some situations, directly influence a plant appraiser's work by directing the appraiser toward specific appraisal approaches and methods.

The Internal Revenue Service. The United States Internal Revenue Service (IRS) offers insight into the issue of reasonableness. IRS Publication 547 – *Casualties, Disasters, and Thefts* comments on regulations pertaining to the appraisal of loss in value (Internal Revenue Service 2015). Casualty claims through the IRS are discussed in Chapter 9.

Industry Standards. The USPAP Scope of Work Rule requires appraisers to produce "credible" appraisals, as follows:

> *The scope of work is acceptable when it meets or exceeds: the expectations of parties who are regularly intended users for similar assignments; and what an appraiser's peers' actions would be in performing the same or a similar assignment.* (The Appraisal Foundation 2016)

This standard may work better for the real estate appraisal profession than for other professions where there is less standardization, accountability, and regulation.

Reporting

As noted in Chapter 3, a report communicates the assignment result and supporting information to the client. Its form depends on the nature and scope of the assignment. The American Society of Consulting Arborists' *A Consultant's Guide to Writing Effective Reports* (Keefer 2004) describes report preparation from initial planning through final production. A companion publication, *Example Reports for Consulting Arborists* (ASCA 2013), provides examples of consulting reports.

Report Types

The Appraisal Foundation's *Uniform Standards of Professional Appraisal Practice* (USPAP 2016, AO-11) provides a useful framework for appraisal analysis (Standard 1) and reporting (Standard 2). It describes two types of written report: appraisal report and restricted appraisal report (Table 8.1). An appraisal report describes background information and sets forth the analysis and conclusions in sufficient detail to allow the reader to understand the appraisal process. The level of detail will depend on the needs of the client and intended users. A restricted appraisal report is for the client's eyes only. It is a concise summary of results with statements of facts and conclusion, usually lacking in detail about data and analysis. The report described in Example 1 in Chapter 3 may be considered a restricted appraisal report.

Oral reports may also be provided to the client. These carry the exact same professional responsibilities as written reports, and the appraiser should retain sufficient file notes to enable production of a written report if one is requested.

Report Contents

Appraisal reports should clearly articulate the following:

- name of the client and intended users
- intended use of the appraisal
- identification of the asset being appraised, its ownership, its current use, and whether it is real estate or personal property
- definition of the type of value or cost being estimated and the source for the definition
- effective date of appraisal valuation, date of the inspection, and date of the report
- scope of work
- where estimating market value, highest and best use (HBU) of the asset appraised and the property on which it sits
- appraisal approaches, methods, and techniques employed
- where more than one valuation method is used, a well-presented rationale for the reconciliation process
- appraisal result

The appraisal report should also contain a statement of assumptions and limiting conditions which may affect the scope or nature of the appraisal. According to the USPAP, appraisers should identify the following:

- all extraordinary assumptions; that is, assumptions which, if omitted or altered, could have a significant effect on the appraisal value or cost conclusion
- all hypothetical conditions; that is, conditions that may be different from what exists but are supposed for the purpose of analysis

Table 8.1 Types of appraisal reports

Item number	Appraisal report	Restricted appraisal report
1	State the identity of the client and any intended users, by name and type;	State the identity of the client by name or type; and state a prominent use restriction that limits use of the report to the client and warns that the rationale for how the appraiser arrived at the opinions and conclusions set forth in the report may not be understood properly without additional information in the appraiser's work file;
2	State the intended use of the appraisal;	State the intended use of the appraisal;
3	Summarize information sufficient to identify the real estate or personal property involved in the appraisal, including the property characteristics relevant to the assignment;	State information sufficient to identify the real estate or personal property involved in the appraisal;
4	State the property interest appraised;	State the property interest appraised;
5	State the type and definition of value and cite the source of the definition;	State the type of value and cite the source of the definition;
6	State the effective date of the appraisal and the date of the report;	State the effective date of the appraisal and the date of the report;
7	Summarize the scope of work used to develop the appraisal;	State the scope of work used to develop the appraisal;
8	Summarize the information analyzed, the appraisal methods and techniques employed, and the reasoning that supports the analysis, opinions, and conclusions; exclusion of the sales comparison approach, cost approach, or income approach must be explained;	State the appraisal methods and techniques employed; state the value opinion(s) and conclusion(s) reached and reference the work file; exclusion of the sales comparison approach, cost approach, or income approach must be explained;
9	State the use of the property existing as of the date of value and the use of the real estate or personal property reflected in the appraisal;	State the use of the property existing as of the date of value and the use of the real estate or personal property reflected in the appraisal;
10	When an opinion of highest and best use or the appropriate market or market level was developed by the appraiser, summarize the support and rationale for that opinion;	When an opinion of highest and best use or the appropriate market or market level was developed by the appraiser, state that opinion;
11	Clearly and conspicuously state all extraordinary assumptions and hypothetical conditions, and that their use might have affected the assignment results; and	Clearly and conspicuously state all extraordinary assumptions and hypothetical conditions, and that their use might have affected the assignment results; and
12	Include a signed certification in accordance with Standards Rule 2-3 or 8-3.	Include a signed certification in accordance with Standards Rule 2-3 or 8-3.

Source: Uniform Standards of Appraisal Practice. 2016–2017. Appraisal Standards Board. The Appraisal Foundation. Reprinted with permission.

For example, extraordinary assumptions and hypothetical conditions might include the following:

- It would be infeasible to reproduce the lost tree with a substitute of the same species, size, and condition.
- The appraiser had to rely on predamage photographs of the subject property and was unable to personally verify the predamage condition of the tree. The appraiser determines that the condition at the time of damage was…
  - Where data were insufficient or missing, the following steps were taken:
    - Information (e.g., acres, survey plats, plant inventory, preexisting conditions) provided by the client is assumed to be reliable, and the scope of the assignment did not include data verification.
    - No plants of the species/cultivars being appraised are readily available, so substitute species/cultivars are used in the valuation.
    - The appraiser has estimated replacement value and makes no representation as to what the market value of the plant might be.
    - The appraiser had to visit the site two years after the damage occurred and has had no opportunity to verify whether the conditions immediately following the damage subsequently changed. The cost estimate is based on the extraordinary assumption that the condition of the property did not change between the date of damage and the date of inspection.
    - A tree risk assessment has not been included in the scope of this assignment. The client is advised to seek a tree risk assessment before conducting any remedial work.
    - While the property's HBU is for conversion to an industrial site, the analysis is based on the hypothetical condition that the property's HBU is for single-family residential use.

Setting forth assumptions, limiting conditions, and hypothetical conditions provides the appraiser with the opportunity to clearly disclose to the users of the appraisal report what sorts of assumptions, limitations, or caveats frame the valuation and to explain the context within which reasonableness should be measured.

Example Using Multiple Approaches

Combining Component Analysis (Sales Comparison Approach) and the Cost Approach to Estimate Market Value

If it is not practical to derive the value of an individual plant by comparing real estate sales with and without trees, the appraiser can combine the component analysis method with the reproduction cost method to estimate the market value of each component. The appraiser must first estimate the total value of the landscape from a sales comparison analysis. Then the appraiser applies the depreciated reproduction cost methodology to each landscape component and totals these values to estimate the total depreciated reproduction cost. The appraiser divides the cost of each component by the total cost for the entire landscape to provide a percentage value for each component. The total value of the landscape from the sales comparison approach (SCA) is then multiplied by the percentage value of each component to estimate its market value.

For example, in an eminent domain condemnation proceeding for a highway-widening project, an appraiser should estimate the economic loss from the removal of three shade trees and the loss of a driveway, a portion of a lawn, and some shrubs and perennials. The plant appraiser works with a real estate appraiser, who estimates that the entire property is worth $310,000 and the landscaping contributes $15,000 (4.8%). The plant appraiser applies the cost approach to derive the depreciated reproduction cost estimates for each component of the landscape, which is $19,000 in this example. The steps used for developing loss estimates are as follows (Table 8.2 also shows a breakdown in allocating market value to each landscaping component):

Step 1. Estimate reproduction or replacement cost new less deductions for depreciation (omit external limitations) [$19,000]

Step 2. Estimate total landscape/hardscape contribution to property value [$15,000]

Step 3. Compute percent of total cost for each component [Column 3]

Step 4. Apply percentages to contributory value [Column 4]

Step 5. Estimate percentage loss by component [Column 5]

Step 6. Compute loss by component [Column 6]

Step 7. Sum losses [$8,250]

Table 8.2 Allocating market value to component landscaping.

Category	Cost estimate ($)	Cost estimate (%)	Market allocation	Landscape loss (%)	Landscape loss ($)
Lawn	$3,000	15.8%	$2,368	15%	$355
Driveway	$2,000	10.5%	$1,579	100%	$1,579
Patio, retaining wall, fencing	$4,000	21.1%	$3,158	0%	–
Hedges, shrubs, perennial flower beds	$2,500	13.2%	$1,974	50%	$987
Shade trees (5)	$7,500	39.5%	$5,921	90%	$5,329
Total	**$19,000**	**100%**	**$15,000**	**100%**	**$8,250**

The appraiser estimates the amount of loss for each of several landscape components. The highway widening is taking 15% of the lawn area and three of the shade trees from the front yard, closing the entrance to the driveway (100% loss), and removing half (50%) of the shrubs and perennial beds. The three shade trees provide an attractive setting for the front of the house and important shade in the summer, whereas two remaining trees are on the side and back of the house. The plant appraiser therefore concludes that the three trees taken comprise 90% of the total tree value, despite the fact that all five are similar in size and condition.

The estimated reduction in property value due to landscaping damage is $8,250.

In the example, the appraiser calculated a dollar value for each component of the landscape. The appraiser might achieve similar results by using a more intuitive approach of simply assigning a percentage value to each component and multiplying the percentages by the total landscape value ($19,000).

CHAPTER 9

Additional Applications: Wooded and Forested Areas, Trees near Utilities, Historic Trees, and Casualty Loss

CHAPTER OUTLINE

Overview

This chapter discusses additional aspects of plant appraisal such as wooded and forested areas, utility rights-of-way, and historic trees. An example of a timber harvest appraisal is included. Aspects of casualty claims and estimating damages are also discussed.

Wooded and Forested Areas

One of the most perplexing situations for plant appraisers involves wildland areas that are not managed for **timber** and have an alternative, non-consumptive highest and best use. Plant appraisers most commonly encounter this situation around the urban interface.

Where the wooded area encroaches on a developed property and its trees clearly enhance the general appeal and value of the property, then shade tree valuation methods may apply. Damage situations may call for the cost approach (TFT, CCT) or the sales comparison approach. Where the cost approach is applied to native tree cover, the CTLA generally recommends using functional replacement over reproduction cost.

For trees that do not encroach on developed property, the native forest may offer certain aesthetic values, but they come at little cost to the landowner and the contribution to total benefits that individual trees make can be marginal. In such situations, it may be that a case can be made for either dismissing extrapolated cost methods altogether or applying them with due consideration of how typical owners benefit from such properties. One appraiser might concentrate on cost with no regard for market value, whereas another might do the opposite.

As one gets farther into the forest from the urban interface, extrapolated cost methods become more difficult to defend as a legitimate valuation tool.

Example: View Clearance

The owner of a home in a high-end subdivision hires a tree care contractor to remove trees in an adjacent county park in order to enhance the view from her home. The trees in question are part of an unmanaged native forest, ranging in size from saplings to large, overmature specimens. The affected area is approximately one acre, encompassing a relatively small portion of the total area of park.

The county sues the residential owner for trespass and hires Appraiser A. The county wants to establish the cost to reproduce the damaged area as it existed the day before the trees were removed. Appraiser A uses the cost approach, reproduction method, and trunk formula technique to conclude that the value of the destroyed trees was $500,000. Appraiser A explains that no depreciation should be deducted because the trees were "natural," despite obvious structural and health issues.

The defendant (the landowner) hires Appraiser B to develop an estimate of the market value of the trees. Specifically, the appraiser is asked to estimate the extent to which the market value of the county property has been diminished by the act of cutting the trees in question.

Appraiser B's report provides an opinion of the park property's market value. The local tax assessor tells Appraiser B that if the area of cut trees were placed on the open market, it would be bought for residential use similar to the rest of the subdivision. The assessor provides data for adjacent properties and affirms a market value of one acre as approximately $250,000. Appraiser B also hires a local real estate appraiser, who provides a market value of $265,000 for the one acre in its predamage condition. The appraiser concludes that development of the property would remove most of the trees on the one acre. In the local area, the average contribution of landscaping to total land value is 15%. This suggests that the contributory market value of some trees preserved during development of the one acre would be in the range of $37,000 to $39,750 ($250,000 × 0.15 and $265,000 × 0.15) for the entire property. Appraiser B estimates that a few well-placed native trees in a typical landscape should contribute about 30% to the overall landscaping value, or approximately $11,500 ($37,000 × 0.30 and $39,750 × 0.30). Appraiser B concludes that the market value of the removed trees is $11,500.

Appraiser B also notes that the impact of the one-acre removal in the overall context of the park is minor. Aside from the one acre, the park's tree canopy remains intact. No trails pass through the cut area. Views from perspectives other than the adjacent property owner are largely unaffected.

In reviewing Appraiser A's report, Appraiser B notes:

1. The assignment result was represented as a value, but Appraiser A did not provide a definition of value. The trunk formula technique produces a cost estimate. This view was based on the premise that the cost approach resulted in an estimate of the amount that trees contributed to the market value of the land.
2. No depreciation was applied, that is, depreciation was 100%. While TFT does not require depreciation, considerations of tree condition, stand development, and planting density are typical when using this technique.

In reviewing Appraiser B's report, Appraiser A notes:

1. An analysis of HBU provided a market value. Yet the property in question is unlikely to ever be placed on the open market. As a county park, it is owned and managed in the public interest. Trespass and cutting trees to improve a view violated that interest.
2. That the rest of the park remains intact does not diminish the extent of damage for the one acre.

At mediation, the property owner who hired the tree care company acknowledges that the purpose of the removal was to improve a view. The property owner had been told by the realtor who sold the property that a view would increase the market value of the property 30%. County officials acknowledge that use of the park is not affected by the tree removal. They also acknowledge that the County would not spend $500,000 to install and care for large trees in the cut area.

The mediator observes that Appraiser A prepared an estimate of the cost to reproduce the one acre of trees while Appraiser B produced an estimate of the market value of the property and the trees' contribution to it. The mediator observes that Appraiser A's result is twice the market value of an undeveloped property. The mediator questions Appraiser B about market value estimates of the cut area based on HBU. Appraiser B acknowledges that the value of the trespasser's property has increased as a result of the cutting and the resulting view. After questioning the appraisers, the mediator asks Appraiser A to (1) incorporate depreciation into TFT analysis and (2) develop a second estimate using cost compounding reflecting typical forest regeneration methods and costs. The mediator asks Appraiser B to estimate the change in value of the defendant's property, which benefitted by enhanced views. This advances the concept of unjust enrichment, which courts occasionally consider relevant.

The mediator, like a judge, arbitrator, and jury, has to evaluate diverse ideas about how to estimate the loss associated with cutting one acre of woodland. In addition to the testimony of the two appraisers, the mediator considers a wide range of factors, including the details of the situation, the demeanor and attitude of the parties involved, the culture of the region, and legal standards. Similar cases have awarded damages based on stumpage value, cordwood, market value, reproduction cost, and restoration cost. Double and treble damages have been awarded, based on the trespasser's intentions, and such deliberations often meld together valuation metrics and punitive damages. Such decisions can provide ambiguous or misleading guidance to appraisers.

Historical Perspective: Previous Editions of the *Guide*

Previous editions of the *Guide* have viewed trees in landscape settings as distinct from those in forests.

> *Trees growing in unimproved or natural wooded areas have a different value than trees on improved property, such as locations where they have been professionally planted and maintained as a part of the landscape design. Street trees, park trees, and trees in recreational areas have a completely different set of values than those growing on improved residential property.* (Shade Tree Evaluation Committee 1975)

In the third through the seventh editions of the *Guide*, depreciation for location was framed in terms of land use (Table 9.1). Terminology changed from 1975 (third edition) to 1988 (seventh edition). Trees on developed land were consistently rated higher than those on unimproved or undeveloped land. Trees in wooded and forested areas received the lowest rating.

Authors of early editions of the *Guide* were not professional foresters and had little understanding of how timber is appraised. Their perspective was simply that all other things being equal, trees in wooded areas and forests would be less valuable than those in developed areas. This view was based on the context that the cost approach resulted in an estimate of value. Those editions of the *Guide* also recognized that cost methods such as TFT were not appropriate for every situation and that a professional forester should be consulted.

With publication of the eighth edition (1992), depreciation for location shifted from land use to considerations of site, placement, and contribution. Values for these three subcomponents were then averaged. Specific discussion of wooded areas was eliminated.

Some plant appraisers rely on TFT in forest settings, applying low location values (high depreciation). Doing so reflects the general thinking found in early editions of the *Guide*. Others have used TFT without depreciating for location, arguing that trees in natural settings need not be depreciated.

Table 9.1 Determining location values, 3rd through 7th editions of the *Guide*.

Site	% depreciation for location	
	3rd and 4th editions	6th and 7th editions
Feature or historical trees	90% – 100%	—
Average residential	80% – 90%	—
Street and boulevard trees	60% – 80%	50% – 80%
Recreation and picnic area trees	60% – 70%	—
Native open woods trees	30% – 40%	—
Dense forest trees	10% – 20%	—
Park and wildlife preserves	—	40% to 60%
Woods: managed or open	—	20% to 60%
Woods: unmanaged dense forest	—	10% to 30%

Appraisal Considerations for Wooded Areas

Fundamentally, the cost of producing a tree in a production landscape nursery is connected with the value of naturally occurring trees in a forest setting. Where trees are being grown for timber and other wood products, they should be appraised as such.

The ninth edition of the *Guide for Plant Appraisal* (2000) strongly advocates benchmarking assignment results against the market value of the overall property. It also recommends applying common forest regeneration practices where a damaged timber stand can be restored.

Where trees in wooded areas are not being grown for timber, cost approaches such as the trunk formula technique and cost compounding have been successfully employed. Consider two situations:

1. A utility needs to widen its wooded right-of-way in a rural residential area. The utility argues that trees on residential properties are forest remnants left over from the previous working forest and are most appropriately valued as timber or cordwood. The property owner argues that when the land was converted from forest to single-family housing, most of the trees were harvested. Therefore, any remaining trees were converted from timber to landscape.
2. A couple purchases acres of forestland, where they build their retirement home. Some years later, a neighbor logs his land. The logger inadvertently removes several trees from the retirement property. Is this land residential or forest?

How does a plant appraiser identify whether appraisal methods described in this *Guide* are appropriate for a particular situation? The following questions may be helpful:

- What is the nature of the appraisal problem? Is a cost or value desired?
- What is the intended use of the property? Does it differ from the actual use?
- What is the management history of the property? Are the trees naturally occurring or planted? Is a management plan in place? Have any silvicultural treatments been applied?
- Are the trees at issue part of the near landscape or are they in the far landscape or not visible at all?
- What is the intended use of the trees?
- Is the location urban, suburban, rural, or at the interface between urban and rural?
- What functions and benefits do the trees provide? How do people interact with the trees?

Answering these questions should assist a plant appraiser in selecting an approach, method, and technique most appropriate for the setting.

In summary, where trees are part of natural forest cover, the CTLA recommends using methods and techniques that are relevant to the historic, current, and prospective use of the property. Here value is viewed through the eyes of owners, their investment in the property, their use of the property, and what they do to perpetuate their enjoyment and use of the forested property. For this reason, reproduction methods such as trunk formula technique are usually not appropriate. Value is more a function of service potential—or utility—than it is a function of size and the cost to produce a tree in a nursery.

The CTLA also recommends using functional replacement cost over reproduction cost in cases where public benefits are to be restored, as in the example that opens this section. Just as with market value, the CTLA's thinking arises from the principle of substitution. Would any public agency spend several hundred thousand dollars to restore the trees in a natural woodland where a more cost-effective solution is available?

In a litigation setting, it may be appropriate for the appraiser to develop multiple estimates, including the following:

- reproduction cost
- functional replacement cost
- the cost to plant a sapling, with or without any extrapolation
- timber stumpage value
- contributory market value

The appraiser may elect to discuss the results with the client before producing the appraisal report. The results will not always be reconcilable, but they provide a range of outcomes to consider. It may also protect against a court ruling that a particular methodology is appropriate but the appraiser decided not to consider it. The client may elect to have one or more solutions presented in the report.

Appraising Trees in Commercial Forest Settings

Standing trees grown for the production of wood products are called timber. Assignments requiring timber appraisal arise from various situations, such as the following:

- timber investing
- estate planning
- taxation issues (e.g., estate taxes, purchase price allocations, depletion accounting)
- timber sale analysis for buyers and sellers
- casualty loss valuation (federal income taxes), as discussed previously in this chapter
- timber trespass (theft or other illegal harvesting), as discussed previously in the *Guide*
- charitable contributions
- conservation easements
- timber lease and timber deed valuation
- timber valuation

Most of the situations listed above call for estimating either the market value of timber or the contribution that timber makes to the market value of the property of which it is a part. Like shade trees, timber is real estate; therefore, its value is a function of its characteristics and location. Once timber has been cut, it becomes personal property, because it no longer is attached to the land.

Timber differs from shade trees in that it is a severable interest; that is, its value is a function of anticipated harvest revenues after it has been severed from the stump. Therefore, a timber appraisal may require the appraiser to estimate the commercial value of the timber without regard to the value or HBU of the underlying land. This may seem like a violation of the principle of consistent use. However, if the assignment calls for estimating the market value of standing timber, it really calls for appraising a specific legal interest in the property: the right to harvest and sell trees.

The Appraisal Subcommittee of the Federal Financial Institutions Examination Council, which is responsible for overseeing the regulation of real estate appraisal in the United States, characterizes both timber and minerals as severable interests. Because special expertise is required to analyze the markets unique to such interests, timber and minerals are in certain instances exempt from real estate appraisal regulations so long as the land is not included in the interest appraised. It is not uncommon, and it is recommended for both plant and real estate appraisers lacking forestry expertise, to engage the services of a professional forester to assist with the timber portion of a valuation assignment. The CTLA regards this as good advice, and it is consistent with guidelines in the USPAP's Competency Rule.

In the case of timber trespass assignments, where the landowner's trees have been illegally cut by another party, it is not uncommon to require multiple types of analysis, including market value, reproduction cost, functional replacement cost, and statutory penalties based on the tree size and prescribed rates (usually based on tree diameter).

Experts are often called on to appraise the stumpage value of standing trees (i.e., on the stump). Unit stumpage values (often called stumpage rates) are generally expressed in terms of either volume or weight:

- dollars per cord
- dollars per thousand board feet
- dollars per cubic meter
- dollars per green ton

The stumpage value of a tree is therefore a function of how much volume or weight can be realized from the tree, the cost to get the tree to market, and what the most lucrative market will pay for it. Markets range from sawmills, pulp mills, and paper mills to oriented strand board plants, chip plants, biomass-generation plants, pellet mills, and fuelwood producers. Specialty mills like shingle and shake mills, fencing mills, and other forest-product facilities also procure timber products. In some cases, timber brokers serve as wholesale middlemen, buying stumpage or delivered products and remerchandising them to assorted mills.

It takes labor, equipment, and money to convert stumpage to a mill-delivered product. The combined costs to get timber from the stump to a mill is often referred to as the cut-and-haul cost. Loggers must cut the standing timber, skid or yard it to a landing, and process and sort it into various piles according to its size and grade. Once timber is sorted, it must be loaded onto trucks and transported to suitable mills. Transport by rail, water, and ship sometimes augments or supplants trucking.

Some timber is transported in mixed or sorted tree-length loads to a concentration yard, where it is then processed and sorted into various products before being transported to appropriate mills. Hauling costs are directly proportional to hauling distances to mills. So a mill paying $700 per MBF (thousand board feet) located 120 miles away from the subject property may be a less profitable destination for the landowner than a mill paying $650 per MBF located only 20 miles away.

Just as the distance to mills affects stumpage value, so also does the location of the timber on the owner's property. Timber that must be cut from steep slopes using expensive logging methods like cable or helicopter logging will be worth less than similar-quality timber on land that can be logged using conventional ground-based methods (e.g., skidders, mechanized processors). Properties with good access often have lower logging costs and thus higher stumpage values. Timber that is remote and requires long skidding distances will be worth less than timber close to a road. Therefore, location is an important factor in valuing stumpage. The appraiser must account for logging and transportation costs when estimating timber values.

The quality of the timber itself is also critical when estimating the value of merchantable timber, or trees large enough to be sold to a mill. The appraiser typically considers several factors, including species, size (i.e., length, diameter at the small end), and product (e.g., fuelwood, pulpwood, sawtimber). In the case of timber suitable for sawtimber and veneer, the appraiser may also consider product grade. Grading conventions vary by species and region, with higher grades having fewer defects from limbs or live knots, dead knots, sweep, crook, rot, holes, discoloration from fungus, etc.

Sawtimber trees large enough to produce saw logs and veneer often contain multiple products in the same tree. The tops of trees may contain products accommodating smaller diameters, such as pulpwood and fuelwood.

Top and branches may contribute to biomass. Usually the highest value and largest volume or weight will be in the butt log, where diameter is greatest and the proportions of clear, defect-free wood is highest.

The two most common methods for realizing stumpage income are through stumpage sales and direct marketing. A stumpage sale occurs when the owner sells the right to standing trees. Here, the buyer is responsible for cutting the timber and transporting it to a mill of the buyer's choice. Direct marketing occurs when the owner pays a contractor to do the logging and to transport the logs to mills chosen by the owner. That is, the owner contracts directly with loggers and truckers and contracts separately with selected mills offering the most favorable prices. In this case, the stumpage value of timber is computed as a residual value equal to the mill-delivered price less cut-and-haul costs. Sometimes an entrepreneurial profit margin is also deducted in what is called conversion return analysis. In an efficient market where both stumpage sales and direct marketing occur, either method should theoretically produce similar stumpage values. However, for a variety of reasons, it is often advantageous to choose one method over the other.

For timber to have positive stumpage value, the mill-delivered price must exceed cut-and-haul costs (and entrepreneurial profit margin). Where small quantities of timber are evaluated, or where logging costs or hauling costs are great in proportion to mill-delivered prices, it may prove unprofitable to harvest timber. In this case, the appraiser might conclude that its stumpage value is zero. Alternatively, the appraiser might conclude a positive value. For instance, knowledgeable buyers might anticipate markets improving or timber growing to sufficient size to render it economical to harvest.

It is also the case that young timber, particularly on premerchantable plantations, is bought by investors. Here, stumpage value is a function of the expectation of future growth and sale. Young plantations may also be appraised using cost compounding, beginning with the cost to prepare the site for planting and installing small seedlings. Establishment costs may include herbicides or burning to reduce slash (logging residues) and control competing vegetation, and fertilization immediately following planting. These costs are compounded to the age of the trees being appraised. If fertilization or weed control takes place at age five and the trees are eight years old, these costs will also have to be compounded forward to age eight.

There are at least two other variants of the cost approach when appraising premerchantable plantations. One is to develop establishment costs (age zero) and pulpwood value when first merchantable (e.g., age 15) and then interpolate between these two figures for the age class in question (Vicary 1990). The other is to start with the same beginning and ending figures, solve for the internal rate of return (IRR), and then compound the establishment costs at the IRR to the age class in question.

All three appraisal approaches may be used to appraise timber on plantations having young or mixed-age classes. However, the cost approach does not generally apply to mature trees and natural regeneration. In the United States, timber investors do not normally rely on the cost approach, because it violates the **unit rule**. Where large-scale timber investing occurs, the income approach is generally the most significant to the investor.

Applying the sales comparison approach to a stand of merchantable timber may be as simple as discovering local stumpage rates and applying them to the estimated volume or weight of merchantable timber, allowing for differences in timber quality and distance to market. More sophisticated analysis may require the appraiser to identify and inspect local stumpage sales, collect information on each sale (e.g., unit prices, product and grade distribution, market location, distance to market, logging costs, internal access, topography), correlate these factors with the prices paid, and then adjust the unit sale prices from the comparable sales to reflect whether the qualities of each sale are similar, superior, or inferior to the subject timber. Where timber and land both have values, the timber appraiser can often derive timber allocations from comparable sales and apply the indicated unit rates to both premerchantable and merchantable timber. This is common practice when applying the sales comparison approach to large timber investments.

For large timber investments where the timber is sold with the land or under a timber lease or timber deed, the buyer may anticipate growing and harvesting timber over a number of years or even harvest rotations. Here, the appraiser will often apply the income approach. This requires projecting future prices for various products; developing a model projecting typical management and expected timber growth; and projecting land rent, property taxes, and other factors comprising the cash flow analysis (Vicary 1986). The appraiser then converts the projected net operating income into present value using a market-derived discount rate in a process called capitalization, as discussed in Chapter 6. Complex forest modeling is often used, whereby growth and yield models are integrated with economic variables into a linear programming framework that solves for maximum present value, subject to various constraints intended to recognize certain operational realities.

The discount rate is a market-derived interest rate reflecting the rate of return required to attract investment capital to the asset being appraised. Timber that is less liquid in poor markets, lower in quality, or further from harvest age may be viewed as a higher-risk investment, warranting a higher, risk-adjusted discount rate and producing lower present value. Low-risk timber cash flows should produce higher values. Many forests have mixed conditions. Markets flush with investment capital and short on available timber investment opportunities often show little variation in discount rates across a fairly broad range of risk profiles.

Timber and forestland valuation also encompasses non-timber revenues such as the following:

- maple syrup tap leases
- pine-straw sales
- ecosystem services (e.g., wetlands mitigation credits and carbon credits)
- land rental income from cabin leases and recreational uses
- conservation easement sales
- HBU sales (i.e., sales of non-strategic parcels for higher-priced, non-timber uses)

Land use regulations can have a profound impact on the market value of timber. Examples of these regulations include the following:

- municipal zoning laws that limit the extent of forest harvesting
- statewide regulations (e.g., forest practice acts, shoreland zoning, wildlife habitat designations, wetlands, high mountain areas, areas of cultural significance, and sensitive areas)
- forestry limitations contained in current-use property taxation provisions
- deed restrictions (e.g., wood supply agreements and conservation easements) that determine minimum or maximum harvest levels in specified areas
- federal regulations limiting harvesting (e.g., the Clean Water Act, the Endangered Species Act, insect and disease quarantines)

For example, where a forest practice act requires no-cut buffers around clear-cuts, the forest owner must defer harvest in buffers until green-up requirements within the clear-cut area have been met. Where a wood supply agreement commits minimum volumes of timber to be cut and delivered to a specific mill, the timing of harvest and value of this timber will be set by the terms of the agreement.

These are just some of the considerations that timber appraisers may need to consider. The complexity of timber appraisal will depend on the quality, scale, and location of the subject property.

Appraising commercial timber requires special expertise. It is normally performed by professional foresters. In cases where a plant appraiser without forestry expertise is faced with a timber appraisal problem, it is advisable to engage an experienced forester or timberland appraiser for assistance. Some states have forestry license laws; in these states, it is generally the case that one must be licensed as a forester to appraise timber for someone other than oneself or one's employer. Licensed real estate appraisers may also be allowed to appraise timber in these states, provided they meet USPAP competency requirements. Regardless of whether forestry licensing is required, plant appraisers delving into timber appraisal should be competent or affiliate with an expert who is.

Plant appraisers should be wary of applying shade tree appraisal methods to natural and managed forests. Property owners in these settings do not establish forests by selecting the largest commonly available tree from a local nursery. Reasonable and credible appraisal practice applies valuation methods that reflect typical management practice, whether it is a forest, park, or urban landscape.

Damage estimates may employ all three appraisal approaches. For planted forests, the cost approach applies standard reforestation practices—for example, planting 600 trees per acre at an installed cost of 50 cents per tree, including site preparation, seedling cost, and installation. Additional costs often include herbicide control and fertilization. For natural forests, natural regeneration generally comes at little expense and premerchantable trees will have little or no value. However, the loss of small trees can be measured in terms of how long it will take for them to become merchantable (i.e., using present value calculations).

Trees Along Utility Corridors, Easements, and Rights-of-Way

Utilities and public agencies secure access to land for the purpose of transmitting and distributing water, hydrocarbons, electrical energy, and telecommunication services. Facilities such as pipelines, poles, towers, wire, or cable is normally owned by a utility or agency, but the land rarely is. More commonly, both above- and belowground facilities are located within easements. The location of the facility is commonly referred to as a **right-of-way** (ROW) or utility corridor.

Appraisal of plants within and adjacent to utility rights-of-way requires an understanding of the rights and responsibilities of the property owner, the utility, and their contractors, as well as knowledge of appraisal and arboriculture. Although utility corridors are commonly called rights-of-way, they can also be established by easement, prescriptive rights, or be owned by the utility.

Easements

An **easement** is the right to use of another person's or entity's real property for a specific purpose. An easement does not convey ownership or a right to possess or sell the land.

One of the basic functions of an easement is to permit access to another person's or entity's land. Easements may be private (restricted to the easement holder or particular individuals) or public (unrestricted and open to the general public). Easement rights are also created to allow access to sunlight or water; allow views from nearby property; allow support of adjoining land (e.g., retaining walls) or building foundations; allow construction or maintenance of facilities on another person's land; or conserve or preserve the appearance of natural features or the existing use of another person's land or buildings. In the case of utilities, the easement provides the right to construct, operate, and maintain the facility.

Most easements are affirmative; that is, they give the easement holder the right to enter or use the property. Other easements are negative in that they control or limit certain activities. As examples, a view easement might prevent an owner from allowing trees to grow and obstruct a view while a conservation or scenic easement might prevent an owner from cutting trees or altering the natural character of the land. Conservation easements are typically designed to preserve the natural character of the property.

Easements are generally difficult to terminate and typically last forever. However, there are exceptions. For example, temporary and construction easements (or temporary takings in an **eminent domain** context) have an explicitly limited term, and some conservation easements have a defined term, after which they terminate.

Most of the work that utility crews or utility contractors do is within an easement. They have the express right to prune or remove trees that will affect their facilities and to maintain a minimum vegetation clearance. The right to perform vegetation maintenance work, including tree removals, may extend beyond the defined ROW. Utilities may have the right to mitigate risk to facilities posed by off-ROW trees, whether along a city street or in a wooded area.

Rights-of-Way

A ROW is one type of easement, typically a land use agreement that allows passage from one point to another. Common examples of rights-of-way include a private ROW that allows access to a landlocked parcel across an adjoining parcel or a public ROW that allows access to a beach across a private parcel. Public streets and sidewalks are some of the most common ROWs. These are generally owned by the public, which has the right to travel over them. In some states, however, adjoining private parcels may own to the middle of the street and grant easements to the public to travel over them. Utilities may have easements within specific portions of these ROWs.

Utility ROWs include electric, communication, or pipeline corridors. These may comprise easements either on private parcels or those owned by the utility.

Appraisal Considerations for Easements and Rights-of-Way

When working on utility easements and/or ROWs, the appraiser should understand the rights and responsibilities of both the landowner and the utility. Utilities are responsible only for the management of the tree as it relates to the energized conductor or related facilities. Specifications for clearance and system safety define the scope of tree management. For these reasons, line clearance operations may result in a tree that is less than ideal in health, structural integrity, or form. Utilities and their contractors are obligated to meet contract specifications for minimum vegetation clearance. For these reasons, the tree appraiser should understand the contract specifications as well as applicable regulatory requirements in order to determine what is reasonable and appropriate for each situation.

How does a plant appraiser identify whether appraisal methods described in this *Guide* are appropriate for a particular situation? The following questions may be helpful:

- What are the rights and obligations of the utility within the specified easement?
- What is the line voltage carried?
- What are the applicable industry standards and best management practices? Are they in the contract? Are they applied in the field?
- Are the trees at issue part of the near landscape or are they in the far landscape or not visible at all?
- What is the management history of the easement?
- Were weather or any site-specific features relevant?

Historic Trees and Landscapes

A **heritage tree** is deemed significant because of its size, form, shape, age, rarity, association with a historic event, or other feature. Other terms applied to trees are *ancient*, *veteran*, *landmark*, *legacy*, *specimen tree*, and *historic*. In this discussion, *historic* encompasses all of the above. Regulations that designate a historic plant or landscape will vary widely from one location to another and from one public agency to another.

Citizens, groups, and agencies seek to maintain their cultural resources by preserving historic properties or trees. Historic status can be applied to buildings, neighborhoods, and landscapes. Historic landscapes may have been designed by a historic figure (e.g., Thomas Jefferson, Frederick Olmsted) or may be naturally occurring. The landscape can be a key component of a site's designation as historic.

The act of designating a building, site, or landscape as historic is not an appraisal function. Moreover, with plant material, features such as extraordinary size or age do not necessarily indicate or confer historic status. The official designation is generally conferred to sites and landscaping that meet certain criteria. For example, the National Park Service (NPS) criteria for granting historic status to sites and landscaping include the following:

- association with a historic figure, event, or use
- age related to the site's history
- botanical rarity or unique characteristics (e.g., exceptional size)
- design effect related to the site
- ecological significance related to habitat or endangered status
- site-specific considerations such as "signature" or memorial plants where signature means a plant or landscape that is unique and may incorporate some of the other criteria in a particularly striking way

The evaluation of landscapes for historical significance can be performed using NPS guidelines. The product of this evaluation is typically a cultural landscape report (CLR), which ordinarily identifies, inventories, and catalogs landscapes. The process of preparing a CLR is extensive and requires special expertise in historic preservation. The appraiser may use this as the foundation for understanding the significance of the asset being appraised.

Not all jurisdictions have local agencies or commissions for historic preservation, but all states have a state historic preservation office that coordinates activities across the state. At the federal level, the NPS administers the National Register of Historic Places, the highest designation of historic status in the United States. At the global level, the United Nations Educational, Scientific, and Cultural Organization (UNESCO) designates World Heritage Sites (e.g., Yellowstone National Park, Mammoth Cave National Park, etc.). Each of these levels carries unique criteria for its review process. This typically involves a thorough review and documentation of the site, including its specific history, inventory of current conditions, and plans for preservation.

Appraisers should recognize situations where landscapes or plants may be historic. These situations include historic districts, historic sites, or sites that may fit other relevant criteria above. Generally, historic designation increases the market value of sites, buildings, and landscapes. However, designation often limits legal uses or requires additional expenditures, which may reduce the site's market value.

The value of, or valuation methods for, appraising historic or heritage plants may be set by statute, ordinance, or other binding language, such as an insurance policy. In such cases, the appraiser should comply with relevant provisions.

Estimates of the values of historic trees and landscapes can be prepared using the cost, sales, or income approaches. In some situations, it may be appropriate to recognize premiums for historic landscapes or plants, reflecting the excess over depreciated cost or market value that such assets might be worth to the public. The CTLA, however, does not support adding arbitrary or unsupported premiums. Appraisal approaches addressing non-market public-interest value require special valuation techniques like **contingent valuation** and are beyond the scope of this publication.

Casualty Claims and Damage Loss

A **casualty loss** is the damage, destruction, or loss of property due to an event that is sudden, unexpected, or unusual (Internal Revenue Service 2015). The event must be atypical, that is, not a common occurrence. This includes earthquakes, hurricanes, tornadoes, floods, storms, volcanic eruptions, shipwrecks, mine cave-ins, sonic booms, vandalism, fire, smog, and car accidents that were not caused by the owner's willful negligence or action.

Owners may receive partial or full remuneration for the value of plants that were casualties. Three methods for recovery are (1) insurance coverage, (2) income-tax deductions, and (3) civil damage claims.

Trees, shrubs, and other plants on a residential property that are damaged or destroyed by a casualty may be a deductible casualty loss. In order to claim a federal income tax deduction, a decrease in the total value of the real estate must be established.

The following information is intended to be a general overview. The examples of law and insurance coverage pertain only to the United States. Laws and insurance coverage regulations are constantly changing. The client should be advised to consult the appropriate authority for guidance.

The term *sudden loss* is ambiguous, especially in the horticultural world, and may include winter injury, rodent girdling, lightning damage, freeze damage, and certain insect and disease attacks. Typical exceptions to damage losses include the following:

- Insects and diseases. The progressive damage or destruction of trees, shrubs, or other plants by fungi, diseases, insects, worms, or similar pests is not a deductible casualty loss. A sudden destruction due to an unexpected or unusual infestation of insects or disease, however, may result in a casualty loss.
- Progressive deterioration and drought. Because drought causes damage through progressive deterioration, it is not considered a sudden event. Further, damage by drought can usually be controlled by providing water through artificial means.
- Loss of future profits. For example, ice-storm damages to standing timber that reduce the rate of growth or the quality of future timber are not deductible. To qualify as a casualty, existing timber must be rendered unfit for use.

Insurance Coverage

Many residential (homeowners) insurance carriers use a standard insurance form developed by the Insurance Services Office (ISO, a subsidiary of Verisk Analytics, Inc.). The ISO standard form, HO 3, states that trees, shrubs, and lawns are covered for loss caused by fire, lightning, explosion, a riot or civil commotion, aircraft, vehicles not owned or operated by the resident, vandalism, malicious mischief, and theft. There are limits to insurance (liability) for all trees, shrubs, plants, and lawns. This is intended to cover the removal and replacement of the plant. Some insurance companies may provide additional coverage, or a separate policy, for individual trees or plants covering all risks (including wind) and with limits much higher than what is included in ISO's HO 3 form.

In general, damage to trees from wind, tornado, or hurricane is not covered unless the tree damages the home or other insured structure (e.g., shed, deck, fence). In such cases, insurance policies generally cover only the cost to remove the tree from the structure and the hauling away of debris. There is generally no limit to what policies will cover to remove the tree from a structure. Debris removal from the yard, however, is generally limited to $500 or $1000. Clients should check their individual insurance policies.

Plants on commercial properties (including nurseries, orchards, etc.) are covered under a different form, and those policies may provide more comprehensive coverage. For example, golf course policies regularly provide coverage to replace destroyed trees at per-tree limits of $5,000 to $10,000 or higher.

Civil and Criminal Damage Claims

If the damage cannot be satisfied through routine insurance coverage, the plaintiff has recourse through the courts, at which time the services of an attorney, as well as a plant appraiser, are usually needed. A civil or criminal damage claim is a legal means by which a person can take action to recover losses resulting from negligence or intentional or unintentional acts.

Double and Treble Damage

Double damage and *treble damage* are legal terms that describe the amount one may claim for injury to persons or damage to property, typically as a result of a tort—a wrongful act or failure to act. The amount of this **punitive damage** varies from state to state. A fixed multiplier is established by law for certain torts and provides that the claimant who is awarded damages may receive twice (or three times) the amount of the value of the damages awarded. The appraiser, therefore, must find the actual value of the damage. The appraiser should not suggest double or treble damages in his or her appraisal; this is the responsibility of the attorney or court.

Income Tax Deductions

This section deals with federal income tax rules set forth by the U.S. Treasury Department and administered by the Internal Revenue Service (IRS), which adds its own rules, regulations, and letter rulings. Tree valuation issues arise when taxpayers seek to file casualty loss claims for federal income tax deduction purposes.

In this chapter, note that the term *casualty loss* is used in a very specific way, and not to refer to the general concept of loss damage. For the purposes of this section, casualty loss is defined as "the damage, destruction, or loss of property resulting from an identifiable event that is sudden, unexpected, or unusual. It must be swift, not gradual or progressive; ordinarily unanticipated and unintended; and atypical, not day to day" (Internal Revenue Service 2015).

The IRS generally requires valuation supporting casualty loss claims to be based on before-and-after analysis. The appraiser must measure the extent to which the fair market value of the property has declined as a direct result of the casualty event. This is referred to as the diminution in value. To understand how this applies, it is imperative to understand that the property is, generally speaking, the unit of real property (including any improvements such as a house) for which a cost basis has been established.

For damaged trees on residential property, the appraiser measures the value of the entire residential property before and after the event. For a natural or planted **forest**, the appraiser would measure the value of the entire property, not just the specific area affected. In either case, the taxpayer must be able to demonstrate what the adjusted cost basis is for the trees on the property. This is the original acquisition cost, adjusted upward for capital additions (e.g., planting costs) and downward for depletion (analogous to depreciation) from removal or losses.

The taxpayer may have a record of what portion of the acquisition cost of the property was allocated to the trees and subsequent adjustments to this cost basis (downward as trees were removed or lost, and upward as trees were regenerated or planted). If the taxpayer has no record of what portion of the total property's cost basis is in trees, the IRS may allow the taxpayer to establish this cost basis after the casualty event. The plant appraiser may play a role in this as well.

IRS regulations refer to the entire property as the single identifiable property (SIP) that is damaged or destroyed. This is usually straightforward where shade trees are concerned but is more ambiguous in forestry situations. Several prominent legal cases have addressed the matter (Westvaco v. Commissioner, International Paper v. Commissioner), resulting in today's interpretation of the SIP to mean the entire **depletion block** (IRS Rev. Rul. 99-56), which effectively allows the taxpayer to maximize the basis limitation. The depletion block is the unit of property used for purposes of determining the depletion unit, or rate at which timber cost basis is written off when timber harvests occur. The depletion unit is simply the adjusted cost basis in the timber divided by the volume (e.g., cords, board feet, or cubic meters) or weight (e.g., green tons). For large forest holdings, the depletion block may encompass multiple parcels covering a total of 100,000 acres or more, particularly in the case of a forest-industry owner who relies on a particular land base to furnish raw material to a mill.

Casualty loss claims are limited to the lesser of the diminution in value of the SIP or the adjusted cost basis of the trees in the SIP. In most cases, only a portion of the trees are damaged or destroyed by a casualty event. Therefore, the smaller the area affected is in relation to the entire SIP, the more likely it is that the loss in market value will not exceed the adjusted cost basis. For instance, suppose that the cost basis of trees in a SIP is \$100,000; the casualty event destroyed 10% of the trees, with a cost basis of \$10,000; and the diminution in value is \$50,000.

The casualty deduction is \$50,000, the lesser of \$100,000 and \$50,000. IRS foresters represent this as borrowing basis. That is, even though the cost basis in trees destroyed is only \$10,000, the taxpayer may write off \$50,000. In the course of harvesting timber, its cost basis is written off as it is harvested. But here, the taxpayer gets to write off not only the cost basis of trees damaged or destroyed, but also an additional \$40,000 of cost basis represented by trees that were not damaged or destroyed.

Another advantage of the casualty loss provisions is that the taxpayer may apply the loss against ordinary income; that is, the taxpayer may write it off at the ordinary income rate, not the lower capital gains rate, even though timber is a capital asset and related income is normally taxed at the capital gains rate.

A key principle in casualty loss valuation—and this applies to federal and state acquisitions as well—is the unit rule, which holds that:

> *Different physical elements or components of a tract of land (such as the value of timber and the value of minerals on the same land, irrigated cropland and dry cropland on the same parcel, etc.) are not to be separately valued and added together.* (Appraisal Institute 2010)

One application of the unit rule is reflected in the requirement that the valuation analysis must apply to the market value of the entire property, not just the component that is damaged or destroyed.

The Uniform Appraisal Standards for Federal Land Acquisition (UASFLA) elaborates on the unit rule. For example, the value of timber, as an independent component, cannot be added to the value of minerals in the same property, also an independent component, and this sum further adds to the value of the land. Such a procedure results in a summation or cumulative appraisal, which is forbidden in appraisals of federal acquisitions, as it is in general real estate appraisal practice. The summation appraisal is an invalid procedure because the entire unit is being hypothetically sold in its entirety, not as separate individual property. When discussing the separate elements of the property in their analyses in the appraisal reports, appraisers should always clearly state that these elements were considered with respect to their enhancement of the value of a whole (Interagency Land Acquisition Conference 2000).

The unit rule is the basis for the IRS rejecting valuations that view separate valuations of portions of real estate in isolation from the value of the whole. The IRS requires claims to be reasonable with respect to the amount that the damaged asset contributes to the value of the whole. For this reason, the IRS often looks askance at the cost approach, and it is particularly skeptical of formulaic analyses, such as the trunk formula technique.

Because the IRS looks at the value of the entire property, it generally does not accept the trunk formula technique (or the trunk formula method, as the IRS refers to it) or the use of any formulaic methodology in measuring the value of a loss. The preferred method is the sales comparison approach, and the income approach is also accepted for properties bought and sold based on their ability to generate net income.

Casualty damage losses and their reporting must follow specific IRS criteria and standards. Determination of property use (i.e., for personal or business use or as income-producing property) must be established. IRS Publication 547, *Casualties, Disasters, and Thefts* discusses rules for different situations that taxpayers encounter. The IRS has published additional documents, letter rulings, and field directives covering specific situations.

For federally declared disaster areas, the IRS has assembled disaster loss kits containing the publications, forms, and instructions needed to help both individuals and businesses to claim deductions for casualty losses within those designated areas.

Generally speaking, when estimating casualty losses, the entire property—including any improvements, such as buildings, trees, and shrubs—is treated as one item. This means that property owners cannot submit landscape loss separate from loss to their home or from any other improvements to the property. All of the property damage incurred by the owner must be treated as an overall loss to the real estate. This overall loss is determined by using the smaller of (1) the adjusted cost basis of the SIP or (2) the entire property's decrease in fair market value (FMV).

As mentioned previously, the adjusted cost basis of the property is the initial monetary investment in the property, plus any improvements, minus any depreciation. For instance, if the property owner purchases a property for $75,000, makes improvements equaling $35,000, and shows depreciation of the improvements of $10,000, the adjusted cost basis, or book value, is $100,000.

The Department of Treasury defines fair market value as "the price at which the property would change hands between a willing buyer and willing seller, neither being under any compulsion to buy or sell and both having reasonable knowledge of relevant facts" (26 C.F.R. sec. 20.2031-1[b]).

This is essentially synonymous with other common definitions of market value. In Publication 547, the IRS explains that "The decrease in FMV used to figure the amount of a casualty or theft loss is the difference between the property's fair market value immediately before and immediately after the casualty or theft" (Internal Revenue Service 2015). For instance, if the overall FMV immediately before the casualty was $150,000 and the overall FMV immediately after the casualty was $60,000, the decrease in FMV is $90,000.

In lieu of reporting before and after values, the taxpayer may use the actual cost of cleaning up, making repairs, and restoring landscape to its original conditions as a measure of the decrease in FMV by documenting the amount of money spent for

- removing destroyed or damaged trees and shrubs, minus any **salvage value** received for those trees and shrubs;
- pruning and other measures taken to preserve damaged trees and shrubs; and
- any replanting necessary to restore the property to its approximate value before the casualty.

According to IRS rulings 66-242 and 68-29, loss cannot be hypothetical, and values of individual shade or ornamental trees cannot be computed by the use of a formula. Section 1.165-7(a)(2)(ii) of the regulations provides that the cost of repairs to damaged property is acceptable as evidence of the loss of value if the taxpayer shows all the following:

- The repairs were actually made.
- The repairs were necessary to restore the property to its precasualty condition.
- The amount spent for such repairs was not excessive.
- The repairs did not provide more than the actual damage.
- After the repairs, the value of the property, as a result of the repairs, does not exceed the value it had immediately before the casualty.

The standard for judging whether repairs are excessive is based upon their cost in relation to the value of the SIP. Accordingly, the IRS looks askance at valuation practices that are either theoretical or hypothetical, not market based, and violate the unit rule.

The taxpayer filing for a casualty loss deduction must provide proof or evidence of all the following conditions:

- the type of casualty (e.g., fire, tornado, hurricane, theft) and when it occurred;
- that the casualty was sudden, unexpected, or unusual;
- that the loss was a direct result of the casualty;
- that he or she owns the property; and
- any insurance claims made for reimbursement, for which there is a reasonable expectation of recovery.

The appraiser should visit the site to document that the casualty event occurred and that efforts to repair or restore the property were undertaken. The burden of proof for documenting all of the above conditions is on the property owner, and the appraiser can be very helpful in this regard.

As an alternative to documenting clean-up, repair, and restoration efforts, the landowner may submit a property appraisal documenting the difference between the FMV immediately before the casualty event and immediately after the event.

Regardless of which option the taxpayer elects, any revenue from salvage must be deducted from any amount claimed for casualty loss purposes. If you encounter a situation that involves the tax code, consult a tax professional.

Casualty Losses Pertaining to Timber

As noted earlier in this chapter, the cost basis of timber can change over time. Initially, it may be established by deriving the relative proportions of the acquisition cost that should be attributed to the land and timber. Once this is established, the timber cost basis is divided by the quantity (volume or weight) to ascertain the depletion unit. When timber is cut, the capital gain is computed as revenue in excess of the cost basis in the timber removed. If the cost basis prior to harvest is \$20,000 for 50 thousand board feet (MBF), then the depletion unit is \$400 per MBF (\$20,000 ÷ 50 MBF). If the taxpayer harvests 10 MBF, the cost basis is reduced (or written off) by \$4,000 (10 × \$400). The adjusted cost basis is now \$16,000 and the volume is 40 MBF, and the depletion unit is still \$400 per MBF. If in ten years the timber has grown to 60 MBF, the depletion unit becomes \$267 per MBF (\$16,000 ÷ 60 MBF).

As timber grows from premerchantable size to merchantable size, the cost basis in the premerchantable account shifts from one subaccount to the other. Planting and other establishment costs are capital additions and increase the premerchantable account.

If timber can be salvaged after the casualty, the amount received from salvaged timber is treated as a sale, along with insurance proceeds. They both mitigate the amount of the claim and can even result in a capital gain instead of a loss if proceeds exceed the cost basis.

The burden of proof to thoroughly support a casualty loss claim rests squarely on the shoulders of the taxpayer. For extensive timber losses, the IRS has recognized that this burden is particularly onerous. To reduce compliance risk, the IRS's Large and Mid-Size Business Division (LMSB) issued its "Field Directive on Timber Casualty Losses" on April 16, 2004. Rather than appraise before and after values for the entire SIP, the taxpayer may elect to appraise just the timber damaged or destroyed.

This field directive provides what the IRS refers to as a safe harbor, whereby the taxpayer can use a less burdensome method to compute the loss and enjoy greater certainty that the IRS will not challenge their claim.

For a quick reference on timber tax laws, the U.S. Forest Service released its *2011 Federal Income Tax on Timber: A Key to Your Most Frequently Asked Questions* (http://www.fs.fed.us/spf/coop/library/taxpubfaqs.pdf).

Proof of Loss

The claimant must establish that a casualty or theft occurred and support the amount taken as a deduction. Sentimental value is not considered when determining the amount of loss; only actual market value is considered. The claimant should be able to document:

1. the type of casualty and when it occurred;
2. that the loss was a direct result of the casualty;
3. that he or she was the owner of the property or contractually liable as a leaseholder to the owner for damage;
4. the cost or other basis of the property as evidenced by a purchase contract, deed, etc. (improvements to the property should be supported by evidence, such as checks or repairs);
5. depreciation allowed or allowable, if any;
6. fair market value before and after the casualty; and
7. the amount of insurance or other **compensation** received or recoverable, including the value of repairs, restoration, and clean-up provided without cost by disaster relief agencies or others.

Photographs

Photographs of the property before and after it was damaged are helpful in establishing its condition and value before and after the casualty. Photographs showing the condition of the property after it was repaired, restored, or replaced may also be helpful. Each photograph should include the name of the photographer, date taken, location, and photo log number.

Appraisal Procedure for Casualty Loss

When dealing with casualty loss claims, plant appraisers should follow this procedure:

1. Advise the property owners of present rules and regulations and advise them to retain professional assistance.
2. Examine the tree and site, take photographs and measurements, and collect relevant data.
3. Coordinate the plant appraisal report with the property owner and/or with the real estate appraiser; ensure that this joint effort is detailed and accurate.
4. Prepare and submit this report according to standards set by the appropriate consulting industry.

Be prepared. Retain necessary backup material to support the appraisal under the scrutiny of the courts or the IRS.

Example of Additional Applications

Timber Harvest

Mrs. Smith decides to sell the timber on her 20-acre woodlot. She contacts her arborist, who refers her to a forest appraiser. The appraiser samples the woodlot to determine the species, size, and quality of the timber. Results indicate 4 thousand board feet (MBF) of white pine, 11 MBF of hemlock, 10 MBF of red maple, 15 MBF of yellow poplar, and 20 cords of red pine suitable for pulp. The appraiser consults the local stumpage survey (Table 9.2) to determine recent timber sales prices.

Species	**Volume**	**Unit price ($)**	**Value ($)**
White pine	4 MBF	78	312
Hemlock	11 MBF	69	759
Red maple	10 MBF	447	4470
Yellow poplar	15 MBF	205	3075
Red pine	20 cords	7	140
Total			**$8,756**

The value of the timber is $8,756.

Table 9.2 Sample Stumpage Survey – May 2016

Northwest PA May 2016 Doyle Log Rule

	Sawtimber ($/MBF)				Pulpwood ($/Cord)			
Species/Product	**Low**	**Mean**	**Median**	**High**	**Low**	**Mean**	**Median**	**High**
				SOFTWOOD				
White pine	$50	$78	$73	$110	$3	$5	$5	$7
Red pine	$50	$67	$67	$85	$7	$7	$7	$7
Jack pine	$0	$0	$0	$0	$0	$0	$0	$0
Hemlock	$50	$69	$73	$88	$3	$5	$5	$7
Other softwoods	$0	$0	$0	$0	$4	$4	$4	$4
				HARDWOOD				
Sugar maple	$550	$615	$600	$731				
Red maple	$350	$447	$415	$651				
White birch	$73	$73	$73	$73				
Sweet birch	$0	$0	$0	$0				
Yellow birch	$0	$0	$0	$0				
Hickory	$50	$109	$110	$170				
Yellow poplar	$145	$205	$210	$300				
Scarlet oak	$250	$323	$330	$402				
Chestnut oak	$250	$323	$330	$402				
Black oak	$350	$485	$400	$923				
Northern red oak	$600	$697	$655	$923				
White oak	$300	$542	$531	$810				
Mixed oak	$449	$449	$449	$449				
White ash	$350	$499	$490	$727				
Black cherry	$960	$1,190	$1,100	$1,780				
Beech	$25	$43	$48	$50				
Misc. hardwoods	$42	$68	$62	$114	$2.00	$3.49	$3.00	$5.00

Note: Reported log rules vary; International & Scribner converted to Doyle. Sawtimber is woods-run.

Plant Nomenclature

Plant appraisers should learn at least two names for a plant: a common name and a scientific name. There may be several common names for each plant but there is only one unique scientific name. While an Italian stone pine in California is an umbrella pine in Italy, both are *Pinus pinea*.

The system of binomial nomenclature was established by Carolus Linnaeus in *Species Plantarum*, published in 1753 (in Latin). The European and American codes of nomenclature were merged into the International Code of Botanical Nomenclature (ICBM) in 1930. The ICBM was superseded by the International Code of Nomenclature for algae, fungi, and plants in 2011. Cultivated plant names are regulated by the International Code of Nomenclature for Cultivated Plants (ICNCP).

Plant appraisers must learn and use plant nomenclature correctly. The species name consists of two parts: (1) genus and (2) specific epithet, both in Latin or Latinized. The species name for red oak is *Quercus rubra*, where *Quercus* is the genus and *rubra* is the specific epithet. The specific epithet is followed by the authority, the full or abbreviated name of the first person to assign the name to the plant. For example, *Pinus pinea* L. indicates that Linnaeus published the name. In common practice, however, the authority is rarely included with the binomial.

The word *species* is both singular and plural. An undefined species within a genus is written as the genus name followed by sp. (singular) or spp. (plural), no italics.

Example: *Pinus* sp. or *Pinus* spp.

Where the plant is known in the trade by another name, or the plant has been reclassified and the name changed, the incorrect name may be identified as a synonym and placed in parentheses following the correct name.

Example: *Lophostemon confertus* (syn. *Tristania conferta*)

Hybrids between species are designated with an ×. The genus remains the same, followed by × and then the hybrid name. For hybrids between genera, the × is first, followed by the genus and specific names. The × is not italicized.

Examples: *Platanus* × *hispanica* is a hybrid between *P. occidentalis* and *P. orientalis*.

× *Cupressocyparis leylandii* is a hybrid between *Chamaecyparis nootkatensis* and *Hesperocyparis macrocarpa*.

Below the species level are several classifications whose designations depend in part on the whether the plant type occurred naturally (variety), is confined to a geographic region (subspecies), or was cultivated (cultivar) (see table).

The USDA Forest Service produced a list of tree species found in the United States as well as standardized four-letter codes for each. (www.nrs.fs.fed.us/tools/ufore/local-resources/downloads/PDF_UDORE_SPECIES_LIST.pdf)

Table A1.1 Summary of plant nomenclature.

Designation	Description	Notation	Example
Genus (pl. genera)	A group of plants within a family that are morphologically similar and contain one or more species.	First letter capitalized, entire word italicized or underlined (e.g., *Quercus*); may be abbreviated by first initial once written a single time.	*Quercus*
Specific epithet	The second part of a species name.	No capitalization, always used in conjunction with a genus.	*garryana*
Species	A group of plants within a genus having characteristics that are distinct from other groups, and that can interbreed.	No capitalization, entire word italicized or underlined. An undefined species within a genus may be written as the genus name followed by sp. (singular) or spp. (plural), no italics.	*Quercus garryana* *Quercus* sp.; *Quercus* spp.
Subspecies	A group of plants within a species having distinct differences that occur naturally and usually within a specific geographic region.	Lower case, italicized or underlined, preceded by ssp. (abbreviation, not italicized).	*Quercus ilex* ssp. *ballota*
Variety	A group of plants within a species having distinct differences that occur naturally, not limited to one geographic region.	Lower case, italicized or underlined, preceded by var. (abbreviation, not italicized).	*Quercus garryana* var. *breweri*
Forma	A group of plants within a species having distinct variations that occur sporadically and naturally.	Lower case, italicized or underlined, preceded by f. (abbreviation, not italicized).	*Cornus florida* f. *rubra*
Cultivar	A group of cultivated plants within a species having distinct characteristics and which retain those characteristics when reproduced sexually or asexually.	First letter capitalized, not italicized or underlined, in single quotes. If preceded by cv. no quotation marks are used.	*Fraxinus uhdei* 'Majestic Beauty' or *Fraxinus uhdei* cv. Majestic Beauty
Patented cultivar	A cultivar for which a patent has been given. Only the patent holder may commercially propagate or sell cultivar. A cultivar may not be patented and trademarked under the same name.	First letter capitalized, not italicized or underlined, followed by ®; cultivar may be identified by a different name or a string of numbers and/or letters before patent is granted and plant released to trade. Patent number may follow name.	*Betula nigra* 'Cully' The (trademarked) common name is Heritage™ River Birch.
Trademarked cultivar	A name for a cultivar that is trademarked and cannot be used to identify any other similar plant. Plant must also be given a cultivar name approved by the International Cultivar Registration Authorities (ICRA). The trademark may be registered with the U.S. Patent and Trademark Office, and then the ® symbol is used.	First letter capitalized, not italicized or underlined, followed by ™. A trademark may be registered, and then the ® symbol is used.	*Betula nigra* The (trademarked) common name is Dura-Heat™ River Birch.

Note: Adapted from Matheny, N., and J. Clark. 2008. *Municipal Specialist Certification Study Guide.* International Society of Arboriculture. Champaign, IL.

Appendix 2

Calculating Area and Volume

Area

Area is a basic measurement used in most appraisals. It may be an area of land, as in a planting bed or forest tract, the cross-sectional area of a tree trunk, or the projection of crown area. Area calculations are usually based on measurements from rectangles, circles, or ellipses.

Area of a Rectangle

Calculate the area of a rectangle using the following formula:

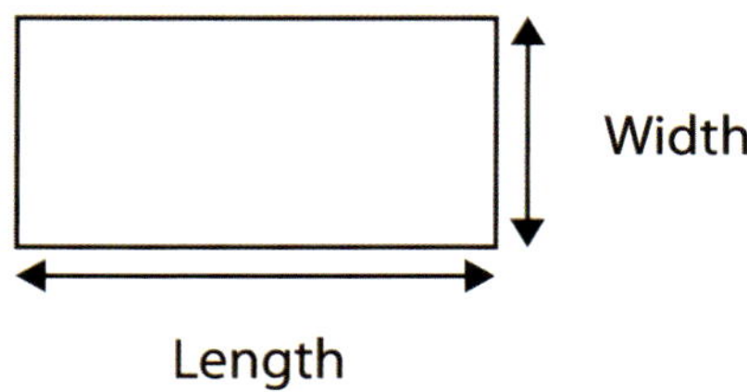

For example, the lawn area in the front yard of a house is 10 feet wide and 50 feet long, so it contains 500 square feet ($10 \times 50 = 500$).

Area of a Circle

Calculate the area of a circle by using either of two formulas:

1. Area = $\pi \times r^2 \approx 3.14 \times r^2$ or $3.14 \times (d \div 2)^2$; or

2. Area = $0.7854 \times d^2$,

where radius (r) is half the diameter (d), and pi (π) is a constant equal to 3.14.

For example, the cross-sectional area of a tree with a diameter of 19.7 inches (50 cm) is 304.7 square inches (1,962.5 cm^2), which can be computed as either $3.14 \times (9.85 \text{ inches})^2$ ($3.14 \times [25 \text{ cm}]^2$) or $0.7854 \times (19.7 \text{ inches})^2$ ($0.7854 \times [50 \text{ cm}]^2$).

Area of an Ellipse

The more a tree's cross section deviates from a circle, the smaller its true area will be for a given circumference. The area of an ellipse will more accurately estimate the cross section of a tree using either of the following formulae:

1. Area = $3.14 \times (R_1 \times R_2)$,

where R_1 and R_2 are the radii of the long and short sides.

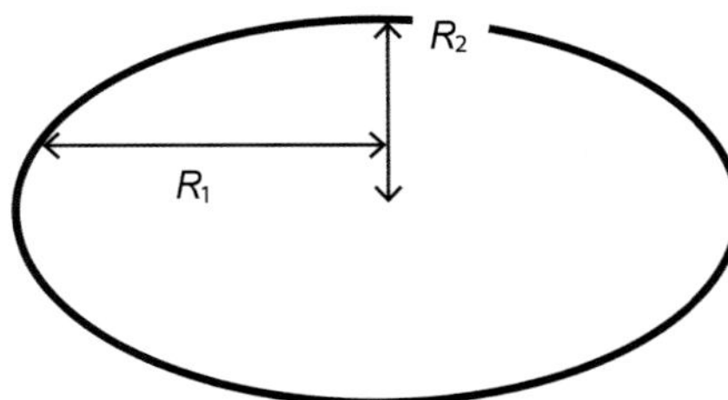

2. Area = $0.7854 \times (D_1 \times D_2)$,

where D_1 and D_2 are the long and short diameters.

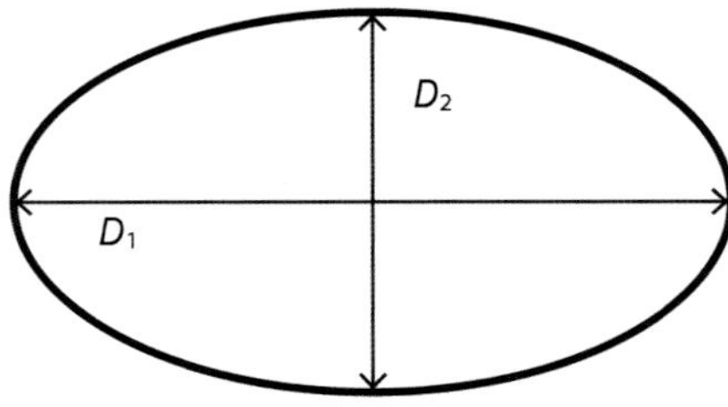

For a tree with diameter measurements of 8.3 and 9.1 inches, these two formulas yield a cross-sectional area of 59.3 square inches.

1. Area = $3.14 \times (4.15 \times 4.55) = 59.3$ in^2

2. Area = $0.785 \times (8.3 \times 9.1) = 59.3$ in^2

Quadratic Mean Diameter (QMD) of Multiple Trees

For a stand of trees, QMD is the square root of the mean of squared diameters for n trees, or:

$$\text{QMD} = \sqrt{(\Sigma\, d_i^2) \div n} = \sqrt{(d_1^2 + d_2^2 + d_3^2 + \ldots d_n^2) \div n}$$

where d_i = the diameter at breast height of an individual tree, and n = the total number of trees measured. The sign Σ represents the summation of a group of data, in this case the sum of the squared diameters for n trees.

For example, suppose there are five trees with diameters of 13, 11, 9, 8, and 9 inches. The sum of squares is 516 square inches and the mean of the summed squares is 103.2 square inches. The QMD is the square root of the mean, or 10.2 inches. The mathematical average is 10 inches.

Volume

Volume is a three-dimensional measurement resulting in cubic inches, cubic feet, or cubic meters. The simplest volume calculation is for a cube:

$$\text{Volume} = \text{width } (w) \times \text{height } (h) \times \text{depth } (d)$$

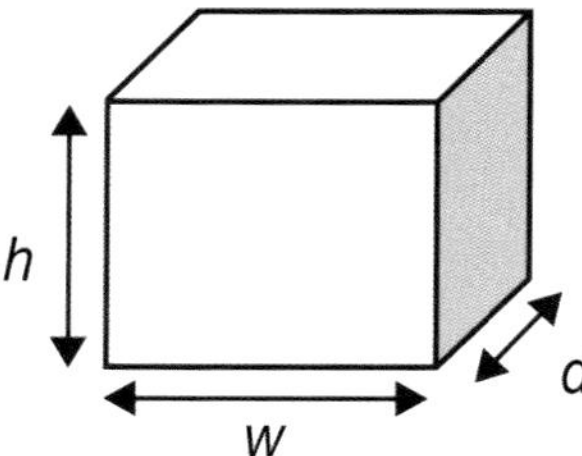

Crown Volume

The footprint of the crown represents the ground shadow it would cast if the sun were directly overhead. If the crown is roughly circular in cross section, its footprint can be estimated by measuring the average diameter of the crown (e.g., by averaging two measurements taken at right angles) and multiplying pi (3.14) by the square of the radius:

$$\text{Area of circle} = \pi \times r^2 \approx 3.14 \times r^2 \text{ or } 0.785 \times d^2$$

When estimating the three-dimensional volume of the crown, techniques will vary according to its shape. It may be helpful to record two width measurements at right angles to each other at the base of the crown and at intermediate points to the top of the tree. Unless highly accurate estimates are required, the appraiser can simply record the average width. Multiply the average widths by crown height to estimate volume. Crown volumes can be calculated based upon simple geometric shapes (Coder 2000). For instance, the volume of a conical crown (most conifers) is

$$1/3 \times (\pi \times r^2 \times h),$$

where r = radius and h = crown height.

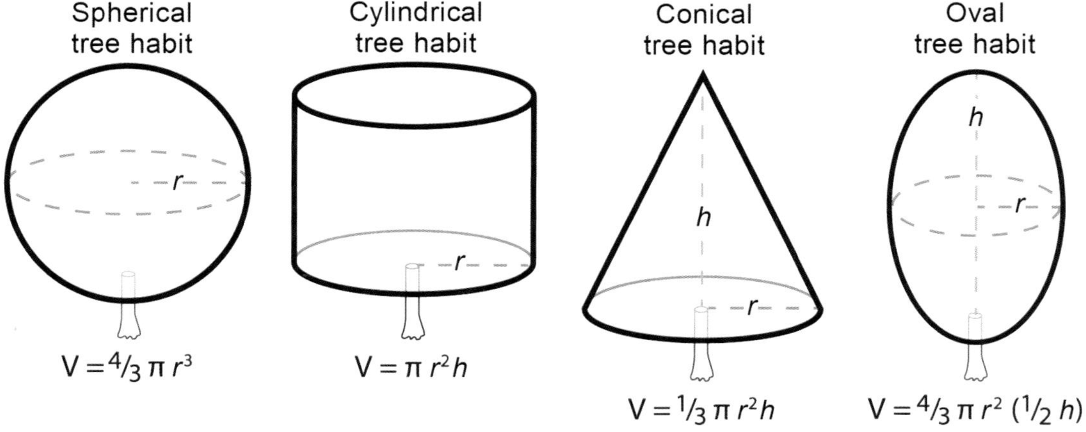

Missing Plants

Appraisers will encounter situations where the subject plants have been completely removed or cut to the ground. Reconstructing the predamage condition requires both quantitative and qualitative data. At a minimum, it is necessary to obtain a count (or estimate) of the number of plants removed and their size. In addition, assessing plant health, structure, form, and planting density will likely be necessary. There are several potential sources of information:

- Photographs and plans from the client can provide information about what the area looked like before the loss.
- Aerial photographs can provide information about the site and plants before the removal.
- If stumps are present, it is usually possible to identify the species, measure the diameter, and assess the health and structure of the stump. Map the stump locations and key reference points, such as property corners, fences, remaining trees, etc.
- Nearby trees (both on- and off-site) can be used as models for plants that were removed. For example, if a property owner removes two street trees located in front of their home, other street trees in the immediate neighborhood may be of the same species that were planted at the same time and managed in a similar manner.

For timber stands, publications provide estimates of log volume based on the diameters of cut stumps. Such tables are useful in situations similar to forested timber settings: unmanaged woodlands, utility rights-of-way, and park preserves. They are less reliable in urban and suburban settings and where trees have been planted.

Data Availability

Some data may be difficult or impossible to obtain. Records may be unavailable or nonexistent. Property owners may not share information or even allow a site inspection. There may be no direct evidence in the form of stumps. For these situations, aerial images and photographs, internet-based mapping, and satellite imagery can be useful.

Where vegetation has been cut, it may be feasible to draw inferences about species, size, distribution, and quality from standing trees that remain on or adjacent to the damaged area. The appraiser should be wary of sampling adjacent areas without first verifying that the sample area is similar in character to the predamage condition of the subject property.

Estimating dbh from Stumps

The diameter of stumps will be larger than dbh because of root flare. It is possible to develop a relationship between stump diameter and dbh by collecting measurements from nearby trees of the same age and species. The accuracy of these estimates is a function of how many sample trees or sample plots are measured, variation in stump heights, and the thoroughness of data acquisition. The more trees of the same size and species, the more reliable the estimate is.

Once a representative population of standing trees has been identified, a sampling protocol must be developed. This may range from systematically measuring as many trees as possible to measuring a sample of trees in selected plots. In either case, the appraiser measures the diameters and cut heights of the subject stumps and then samples standing trees by measuring both diameter at the mean cut stump height and dbh. It is usually not necessary to measure more than 20 trees of the same species with similar diameters to obtain good estimates. As variability increases, the defensible sample size increases.

In the following examples, two ways to estimate the dbh of a missing black oak are shown. More complicated sampling procedures are beyond the scope of the *Guide*. When confronted with situations that require complex sampling procedures, the plant appraiser should follow industry-standard forest inventory sampling and design. For additional information, see Husch et al. (2002).

When collecting tree data on sample plots over a large area, foresters commonly use multiple circular plots of fixed or variable radii to sample hundreds or thousands of trees. Here, it suffices to take a single caliper diameter measurement for each tree (e.g., across the face of the bole facing the plot center) (Husch et al. 2002). With multiple trees, this will produce unbiased diameter measurements.

In situations such as timber trespass, where many thousands of trees have been cut, it may be appropriate to collect sample measurements from hundreds, if not thousands, of sample trees, to obtain sufficient data for predicting dbh across the full range of species and diameters represented. If the harvested area is large enough, it may be impractical to measure all of the stumps, in which case the appraiser measures a sample of the subject trees.

Errors Introduced Through Data Collection and Mathematical Computations

Many plant appraisers and foresters are aware of the pitfalls of data and analyzed data. Appraisers should be familiar with significant figures, measures of central tendency (mean, median, mode, and range), sampling variability, bias, accuracy, and precision.

Example 1. Predicting dbh Using Ratio Analysis

The subject tree was cut at 6 inches (15.2 cm) aboveground and had a diameter of 10.25 inches (26.0 cm) at that height. The dbh can be estimated by measuring the diameter of nearby trees of the same species at both 6 inches and dbh (Table A3.1). A diameter ratio (dbh/diameter at 6 inches) for each sample tree is developed.

Table A3.1 Trunk diameters collected at 6 inches and dbh and resulting dbh/stump diameter ratio.

Tree #	Stump diameter	dbh	dbh/stump diameter
1	12.25	9.00	0.7347
2	8.50	7.00	0.8235
3	14.00	9.75	0.6964
4	10.00	9.00	0.9000
5	12.50	11.00	0.8800
6	10.50	8.50	0.8095
7	12.50	10.25	0.8200
8	9.75	7.75	0.7949
9	12.50	10.25	0.8200
10	11.50	10.00	0.8696
11	9.00	8.00	0.8889
12	12.00	11.25	0.9375
13	8.00	7.00	0.8750
14	11.50	10.50	0.9130
15	10.50	10.00	0.9524
16	15.25	13.75	0.9016
17	10.00	8.50	0.8500
18	8.00	7.00	0.8750
19	12.25	11.00	0.8980
20	12.00	11.00	0.9167
Total	222.50	190.50	--
Mean	**11.125**	**9.525**	**0.857835**

dbh/stump diameter ratio:

Mean ratio*	**0.857835**
Standard deviation	0.0653
Coefficient of variation (SD/Mean)	7.61%

*It is incorrect to divide the sum of dbhs by the sum of stump diameters.

Applying the mean diameter ratio of the sampled trees to the subject tree, the dbh of the subject tree is calculated as:

Predicted dbh = subject stump diameter × mean ratio
= 10.25 inches × 0.8578
= 8.79 inches

Example 2. Predicting dbh Using Regression Analysis

Another technique is to formulate a simple linear regression equation that will predict dbh across a range of stump diameters, including that of the subject tree. Using the 20 trees in the previous example, diameter at 6 inches (the independent variable) is plotted against dbh (the dependent variable). All spreadsheet programs will perform these calculations.

Simple Linear Regression Analysis	
Y-intercept	0.697
Slope	0.794
Regression equation	Y = 0.679 + 0.794 (X)
Predicted *Y* (dbh)	**8.83 inches**
Coefficient of determination (R^2)	0.79

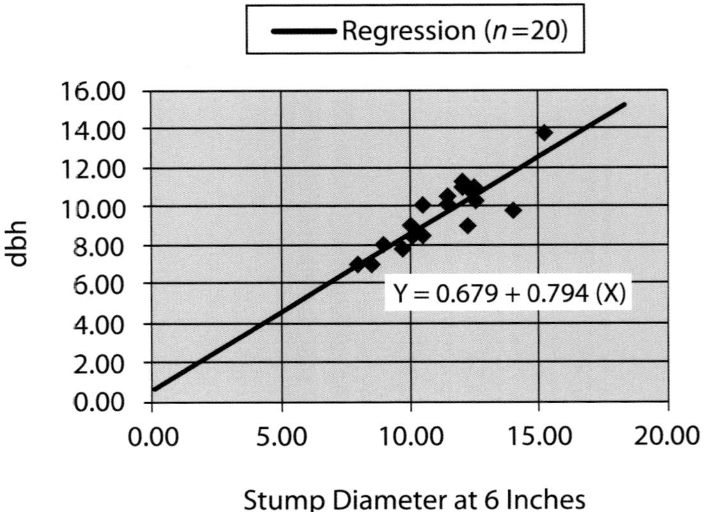

The Y-intercept is the point where the regression line intercepts the Y-axis. The slope is the average change in *Y* (dbh) divided by the average change in *X* (stump diameter). For example, as stump diameter increases by 1 inch, the predicted change in dbh is +0.794 inches. The Y-intercept and slope define the regression equation and, thus, the relationship between stump diameter and dbh. The resulting regression line is the best possible fit describing this relationship, or the fit that results in the lowest sum of squares of the differences between observed and predicted dbh.

The regression equation in Example 2 predicts a slightly larger diameter than the ratio analysis. The coefficient of determination (R^2) measures how closely correlated the independent and dependent variables are to one another. The higher the R^2, the more reliable the regression equation. Technically, the statistic is saying that 78.8% of the variation in dbh is explained by stump diameter.

Identifying the Largest Commonly Available Nursery-Grown Tree

Regional Plant Appraisal Committees (RPACs) are groups of plant appraisers, arborists, landscapers, nursery growers, landscape architects, and university specialists who provide information to tree and landscape appraisers. They collect and organize information that is not readily available to the individual appraiser, including the following:

- species availability;
- nursery stock size and availability;
- regional costs; and
- comprehensive descriptions of plant species.

RPAC membership should be diverse. Where practical, they should have at least one member who is a practicing tree appraiser, arborist, landscaper, nurseryman, landscape architect or designer, and a representative from a university or the Cooperative Extension Service. They should meet regularly to review current recommendations needed for the appraisal process. Updates should be prepared on an annual basis. The information provided by RPACs is most effective when it is up to date.

The largest commonly available transplantable tree has been used in recent editions of the *Guide* as the starting point in establishing the unit cost (dollars per cross-sectional square inch). The eighth edition (1992) noted that "The appraiser must decide what the most commonly available maximum transplantable size will be, depending on species, availability, and the particular site situation" (see p. 53). The ninth edition (2000) suggested that RPACs should determine the size of the "largest commonly available transplantable nursery tree… by the method(s) generally accepted…" (see p. 58).

What method(s) should appraisers or RPACs use to identify the largest commonly available nursery-grown tree? No guidelines or instruction were provided in either the eighth or ninth editions. The phrase itself indicates that it should be a nursery-grown tree and commonly available. The eighth edition suggests that it may vary by species. Size can be a trunk diameter, plant height, crown volume, crown diameter, or container volume.

Per the tenth edition, RPACs should identify "the largest commonly available nursery-grown tree." The following are steps that RPACs might take to do this:

1. Compile information on species availability from nurseries, wholesale suppliers, or garden centers in their region.
2. Calculate the number of species (or taxa) available in each size that trees are sold.

3. Plot the number of species by nursery size, sorted by ultimate tree size.
4. Use the plotted data to determine the largest commonly available nursery-grown tree.

As an example, data on tree availability was acquired from a large wholesale grower on the West Coast. In this region, trees are produced in containers (either boxes or pots) and sold by container size. Species were sorted by mature tree height: small (< 30 feet or 9.14 m), medium (30 to 50 feet or 9.14 to 15.25 m), and large (> 50 feet or 15.25 m) (Table A4.1). The nursery grew 267 taxa in sizes from 15 gallons (56.78 L) to 60-inch (152.4-cm) box.

Table A4.1 Tree species availability by mature tree size class and container size.

Tree size class		No. of taxa	Stock size (container or box)				
(mature height, ft)			15 gal	24 in	36 in	48 in	60 in
Small	< 30	125	112	87	49	12	3
Medium	30 to 50	97	92	74	43	8	0
Large	> 50	45	43	36	24	9	2
Total		**267**	**247**	**197**	**116**	**29**	**5**

Note that species availability declined with increasing stock size: 247 species were available in 15-gallon size but only 29 in the 48-inch-box size (121.9-cm). Additionally, tree height at maturity was not a factor influencing the container size available. Few species were offered in either 48-inch (121.9-cm) or 60-inch-box size. The smaller the stock size, the greater the number of available species.

An RPAC looking at this information could easily conclude that either a 24-inch (70-cm) box or 36-inch box (91.4-cm) represented the largest commonly available nursery-grown tree size. In so doing, their decision would be based on information collected from the appropriate growers.

It is not the largest available nursery-grown tree but the largest commonly available nursery-grown tree that is important. Often the greatest diversity of species is available as seedlings or whips. These tree sizes should not be used for residential or commercial landscapes. Unit prices for tree species that are large at maturity are typically based on nursery trees that range from 1.5- to 4-inch caliper (3.81- to 10.16-cm), rarely as large as 8 inches (20.32 cm).

If truck-mounted tree spades are commonly used in the region, then the largest commonly available nursery-grown tree could be based on this information. Again, the goal is to identify the most commonly available, not the smallest or largest spade in the region.

For wildland and forest settings, local forestry practices should form the basis of RPAC decisions. These include the planting of forest nursery stock (small bare-root or containerized trees), direct seeding, or natural regeneration (coppice, seed trees). For Christmas tree farms, use tree sizes typically used by growers in the region.

This information should be periodically updated by RPACs. If the information is more than a few years old, the appraiser may need to determine the most appropriate size on which to base their estimates.

Compound Interest Calculations

Basic knowledge of compound interest will help the plant appraiser tackle simple and more complex valuation problems. Figure A5.1 presents formulas for computing the present and future values of both single payments and multiple payments (annuities). The payments can be either costs or revenues. Where there are multiple payments, the formulas can be applied only to a regular series of cash flows. If the cash flows are not regular, then spreadsheet applications will be required. In all cases where formulas are used, the appraiser should check the answer using a spreadsheet. Commercial-grade software enables the user to either directly input the compound interest formula or insert a formula containing the appropriate syntax.

Example 1. Projecting Future Plant Care Costs

The cost of tree care increases on an annual basis just like the cost of clothing, food, and housing. From 2010 to 2017 the typical cost inflation rate for tree care in the U.S. Northeast was about 2.0% to 2.5% per year. This was somewhat greater than the relatively low general inflation rate of 1.6%, as measured by the CPI-U, over the same period.[1] In 2017, economists were projecting inflation to increase. Therefore, projecting costs to rise at a nominal rate of 2.5% is reasonable for projection purposes.

Use compound interest formulas to project future costs, just as one would for future values:

$$FC = PC \times (1+i)^n$$

where FC = future cost at age (n), PC = present cost, i = rate of inflation (as a decimal, so 0.025, not 2.5%), and n = the number of years.

This formula is also represented by Formula 1 in Figure A5.1.

If the current cost of a plant heath care visit is \$200 and the service begins immediately and extends another two years, the projected costs will be:

$$FC \text{ Year 1} = \$200.00$$
$$FC \text{ Year 2} = 200 \times (1.025)^1 = \$205.00$$
$$FC \text{ Year 3} = 200 \times (1.025)^2 = \$210.13$$

If the service period is projected to begin a year from now, the projected costs would be \$205.00, \$210.13, and \$215.38 for the three years.

[1]Consumer Price Index – Urban. The average annual rate of increase through 2017 was 2.0% from 2000, and 2.3% from 1990. Annual inflation has ranged from 0.0% to 4.0% since 1990.

Example 2. Calculating the Present Value of Future Costs

Since money can be invested and interest can be earned on that money over time, the option to spend money now on future goods or services means that they should cost less now than in the future. The present value of future costs is computed using the formula:

$$PV = FV \div (1+i)^n$$

where PV = present value, FV = future value at age (n), and i = interest rate.

See also Formula 2, Figure A5.1.

Note that this formula is the mathematic inverse of the formula presented in Example 1. Therefore, we can compute the present value of each of the future costs projected in Example 1, and they should all equal the present cost of $200:

Year 1 = $200.00

Year 2 = 205 ÷ (1.025)^1 = $200.00

Year 3 = 210.13 ÷ (1.025)^2 = $200.00

For reproduction, functional replacement, and repair cost purposes, it may be expedient to compute the present value of the entire series of three payments. If the cost of capital (discount rate) is 6.0%, and the initial payment is immediate, then the future payments are discounted to the present as follows:

PV Year 1 = $200.00

PV Year 2 = 205.00 ÷ (1.06)^1 = $193.40

PV Year 3 = 210.13 ÷ (1.06)^2 = $187.01

Therefore, the total *PV* for aftercare is $580.41. The Excel formula for NPV produces the same answer.

Note that certain formulas in Figure A5.1 enable the plant appraiser to compute the PV for a regular series of payments in one step. For example, if aftercare costs are projected to stay constant at $200 and the first payment occurs at the end of year one, we have a *regular annuity*. Formula 7 results in a PV of $534.60. If we assume the first payment begins at time zero, we have an *advanced annuity*, and Formula 14 yields $566.68. Naturally, the PV of an advanced annuity is higher than for a regular annuity because payments begin immediately.

If we assume that payments increase at the rate of inflation, we can adjust the discount rate of 6% to account for the impact of inflation and simply treat all payments as though they will be $200. It may help to think of inflation *increasing* future payments by 2.5% per year, and the discount rate simultaneously *reducing* their present value at a rate of 6% per year. Formula 17 effectively nets out the two factors by adjusting the discount rate. The 6.0% is a *nominal* discount rate, which means it includes inflation. To convert it to a *real* discount rate, i.e., one that is inflation-adjusted, use Formula 17 to compute: (1.06 ÷ 1.025) - 1 = 0.0341, or 3.41%.

For a regular annuity, where the first payment is $205.00 at the end of year one, we can apply Formula 7 using the real discount rate of 3.41% and simply plug in $200 for *a*. This produces a PV of $561.24. Here, the $200 is increasing at the rate of inflation, but we are netting inflation out of the discount rate, so we assume a flat real payment of $200 per year.

For an advanced series, where the first payment at the beginning of year one is $200.00, we can apply Formula 14 using the real discount rate of 3.41%, again plugging in $200 for *a*. This produces a PV of $580.41.

Conclusion

The formulas in Figure A5.1 are intended to address common situations where it is helpful to compute present and future costs or values using a hand-held calculator. They are also useful for auditing computer solutions. For the appraiser who is inclined to apply compound interest, computer spreadsheets offer the most efficient and accurate solutions, provided the user applies the correct formulas. Moreover, computer spreadsheet solutions provide the only efficient way to address more complex series of irregular payments.

Regardless of how compound interest is applied, the appraiser must understand:

1. How formulas or software functions treat basic algebraic syntax.
2. How to address different timing assumptions, such as begin-period versus end-period payments; whether the first payment should reflect a current or inflated figure; etc.
3. How to derive appropriate inflation and discount rates.

Compound Interest Formulas

Payment Frequency	Direction of carrying values through time: Compounding forward	Discounting backward	
Single	(1) Future value $V_n = V_0 S^n$	(2) Present value $V_0 = \frac{V_n}{S^n}$ (4) Solve for years $n = \frac{\ln(V_n / V_0)}{\ln S}$	(3) Rate earned $i = \sqrt[n]{\frac{V_n}{V_0}} - 1 = \left[\frac{V_n}{V_0}\right]^{\frac{1}{n}} - 1$
	Terminable series	*Terminable series*	*Perpetual series*
Annual (no. of compounding periods = no. of payments)	(5) FV of annuity $V_n = a\left[\frac{S^n - 1}{i}\right]$ (6) Annual payment to yield FV (sinking fund) $a = V_n\left[\frac{i}{S^n - 1}\right]$	(7) PV of annuity $V_0 = a\left[\frac{1 - S^{-n}}{i}\right]$ (8) Annual payment to yield PV (installment payment/ capital recovery) $a = V_0\left[\frac{i}{1 - S^{-n}}\right]$	(9) Capitalization of annuity $V_0 = \frac{a}{i}$
Periodic (no. of compounding periods > no. of payments)	(10) FV of periodic series $V_n = a\left[\frac{S^n - 1}{S^t - 1}\right]$	(11) PV of periodic series $V_0 = a\left[\frac{1 - S^{-n}}{S^t - 1}\right]$	(12) Capitalization of periodic series $V_0 = \frac{a}{S^t - 1}$
Advanced periodic (1st payment is < *t* periods from start of series)	(13) FV of advanced periodic series $V_n = a\left[\frac{S^n - 1}{S^t - 1}\right]S^{t-p}$	(14) PV of advanced periodic series $V_0 = a\left[\frac{1 - S^{-n}}{S^t - 1}\right]S^{t-p}$	(15) Capitalization of advanced periodic series $V_0 = \frac{aS^{t-p}}{S^t - 1}$

Special Cases

(16) Future value of constant dollar change in annual payment (payment increases or decreases starting in period 2) $V_n = g\left[\frac{S^n - ni - 1}{i^2}\right]$

(17) Adjusted percentage rate for geometrically changing payment; first payment = $a(S)$

Compounding forward	Discounting backward	
$i = S(1 + k) - 1$	$i = \left[\frac{S}{1+k}\right] - 1$	When $i = k$, $V_0 = a(n)$

Figure A5.1 Examples of compound interest formulas.

Table A5.1 Compound interest formulas – legend

V	value (V_o = value at time zero; V_n = value at time *n*)
FV	future or ending value
PV	present or beginning value
a	payment made at the end of each payment period (in arears) unless otherwise specified (e.g., in advance)
i	interest rate, expressed as a decimal
S	$1 + i$
n	number of compounding periods
t	number of compounding periods between payments
p	number of compounding periods before first payment is made
g	constant incremental change in dollars of payment, beginning in period two
k	percentage change in payment per compounding period
ln	natural logarithm

Appendix 6

Establishing a Capitalization Rate

As noted in Chapter 6, the capitalization rate (or cap rate) is the ratio of Net Operating Income (NOI) to property asset value. If a property value was $1,000,000 and generated an NOI of $100,000, then the cap rate would be 10%.

Appraisers may use many techniques to estimate an appropriate cap rate. The most dependable starting point is to research past sales of similar businesses. For example, if an appraiser is assigned to appraise Shady Creek Nursery, which has a net operating income of $75,000, then the appraiser will look to prior sales within that region. Research will focus on two primary components: the sale price of each business (presumably their value) and the business's net operating income.

The comparative analysis is shown in the table below.

Business name	Net operating income	Sale price	Cap rate (NOI/sale price)
Smith Nursery	$150,000	$2,100,000	7.1%
Jones Nursery	$85,000	$1,000,000	8.5%
Wilson Nursery	$170,000	$2,300,000	7.3%
Brown Nursery	$45,000	$145,000	31%

How does an appraiser assess this data? A first step would be to discard the Brown Nursery cap rate because it does not appear to reflect typical industry value. Assuming that the subject property's HBU is for continued use as a nursery, the appraiser may reconcile the remaining cap rates by selecting the cap rate of the transaction most similar to the business being appraised, or weighting the three cap rates according to how similar each is to the subject property.

The appraiser evaluates the cap rate from other sales and deems 7.6% as a reasonable estimate. Therefore, the Shady Creek Nursery has an estimated value of $987,000 ($75,000 ÷ 7.6%). This number may not reflect deductions from the value such as outstanding debt, or additive values such as owned real estate, interest income, or capital equipment.

Suppose that the appraiser in the above situation adjusts the cap rate for the Shady Creek Nursery due to a perceived increase in investment risk, such as a small customer base or limited time in the market. After consulting with a qualified expert, the appraiser may add a premium to account for added risk. For example, if the appropriate premium is 1.0%, the adjusted rate equals 8.6% and the resulting value of the business is $872,000 ($75,000 ÷ 8.6%).

Forms

Appraisal Contact Information Form

Client name ______________________________ Date __________ Case # __________

Phone ______________________________ E-mail ______________________________

Address ______________________________

What needs to be appraised? ______________________________

No. of trees _____ Setting: Landscape____ Woodland____ Street trees____ Other ______________

Location in landscape ______________________________

Limits to access ______________________________

Purpose of Appraisal

Reason for appraisal ______________________ Intended use of the appraisal ______________

Other parties involved ______________________________ Date of loss ________

Is legal action expected? Yes ____ No ____

Report type: Oral ____ Letter ____ Form ____ Booklet ____ Due date __________

Site visit: Date ______________ Time ______________ Arborist ______________________

Additional information:

Preliminary Arborist Evaluation

Assignment result: Cost ____ Value ____

Approach(es): Cost ____ Market ____ Income ____

Method(s): Repair ____ Reproduction ____ Func. replace. ____ Other ________________

Technique(s): Direct ____ TFT ____ Cost compound ____ Other ________________

Limiting conditions ______________________________

Time Estimate

Travel ____________ Field ____________ Off-site research ______________________

Analysis ____________ Report ____________ Presentation ____________ Total ________

Fee Estimate ____________

Council of Tree & Landscape Appraisers (CTLA). 2019. *Guide for Plant Appraisal, 10th Edition*. International Society of Arboriculture, Atlanta, GA.

Appraisal Field Data Sheet
Page 1

Client name ______________________________ Date __________ Case # __________

Phone ______________________ E-mail ______________________________

Address ____

Tree Information

Species ______________________________ ID ______________

Plant size: dbh ________@ ________ Height ________ Crown dimensions ______________ Area __________

Adjustments to dbh measurement ______________________________

Crown volume = Height x Base = ______________________________

Age/stage of development: ______________________ Core collected ________

Health ______________________________________ ____%

Structure ______________________________________ ____%

Form ______________________________________ ____%

Comments:

Overall condition ______________________________________ ____%

Placement (use box for sketch/map)

Location of tree ______________________

Distance to reference items ______________________

Soil type, volume, drainage ______________________

Plant density ______________________

Superadequate? ______________________

Topography ______________________

Landscape quality ______________________

Function in Landscape ______________________________

Functional limitations ______________________________ ____%

External limitations ______________________________ ____%

Council of Tree & Landscape Appraisers (CTLA). 2019. *Guide for Plant Appraisal, 10th Edition.* International Society of Arboriculture, Atlanta, GA.

Appraisal Field Data Sheet
Page 2

Current land use ____

Highest and best use ____

Structures/hardscape ____

Management History

Irrigation ____

Fertilization ____

Pruning ____

Pest mgt. ____

Disturbance? ____

Other ____

Damage/Injury

Part affected: Branches ____ Trunk ____ Roots ____ Fruit ____ Other ______________________

Description ____

Date of injury ________________ Extent of damage ________________% Loss ________________

Recovery expected? Yes ____ No ____ Life expectancy ________ yrs

Treatment needed for recovery ____

Estimated predamage: Health ____% Structure ____% Form ____% Overall condition: ____%

Additional Information:

Testing needed: ____ Confirm ID ____ Soil ____ Plant ____ Fruiting body

Glossary

accuracy: (Ch. 4) Correctness; absence of error or bias. An accurate sample represents the true characteristics of the population it is measuring or represents.

accurate: (Ch. 4) Correct in all details; representing the truth.

appraised value: (Ch. 2) An assignment result that identifies the type of cost or value it represents.

appraisal: (Ch. 2) The act or process of developing an opinion of value, cost, or some other specified assignment result.

appraisal process: (Ch. 3) Systematic steps that an appraiser takes to solve for an appraisal problem.

assignment result: (Ch. 2) The appraiser's opinions and conclusions developed for a specific assignment.

assumption: (Ch. 3) A statement or notion taken for granted.

basic cost: (Ch. 5) An estimate of cost before any depreciation is applied.

benefits: (Ch. 2) Attributes that benefit, or provide utility to, someone or something.

capitalization rate (cap rate): (Ch. 6) A percentage rate derived by dividing a property's net operating income by its sale price. A rate divided into the subject property's net operating income for purposes of estimating its market value.

casualty loss: (Ch. 9) The damage, destruction, or loss of property resulting from an identifiable event that is sudden, unexpected, or unusual.

clinometer: (Ch. 4) A handheld optical measuring device used for determining tree height.

compensation: (Ch. 9) The amount awarded by a court to settle a legal claim.

contingent valuation: (Ch. 9) An economic technique for the valuation of non-market resources.

contributory value: (Ch. 2) The amount by which the addition of an item augments or increases the value of the whole, or the amount by which the value of the whole decreases in the absence of that item.

cost: (Ch. 2) The amount of money required to create, produce, or obtain a property or service (Appraisal Institute 2015).

cost compounding technique (CCT): (Ch. 5) A technique for extrapolating current costs to a point in the future, accounting for time in years and applying compound interest. Also known as *cost forwarding*.

cost estimate: (Ch. 2) An assignment result in which the plant appraiser estimates the cost to install, replace, or repair a plant or landscape item.

crop: (Ch. 2) A cultivated plant.

crown: (Ch. 4) The upper part of a tree, measured from the lowest to the highest branch and including all the branches and foliage.

demand: (Ch. 2) The relationship between the quantity of a good desired and factors that affect the willingness and ability of a consumer to purchase the good. The state of being wanted or sought for purchase or use.

depletion block: (Ch. 9) A unit of property used to determine the rate at which timber cost basis is adjusted.

depreciated reproduction cost: (Ch. 5) The cost to replace an improvement with an exact replica, less accrued depreciation.

depreciation: (Ch. 2) A loss in value from any cause; typically caused by either physical, economic, or external factors.

diameter at breast height (dbh): (Ch. 4) A U.S. custom means of expressing a diameter of a tree, as measured 4.5 feet (or 1.37 m) above the ground.

diminution in market value: (Ch. 8) The extent to which the fair market value of a property has declined as a direct result of a casualty event.

direct capitalization: (Ch. 6) An income-based valuation in which the appraiser develops a stabilized net operating income for an asset and an appropriate capitalization rate. The estimated value equals net operating income divided by the capitalization rate.

direct cost: (Ch. 5) The sum of all actual or estimated costs associated with repair, reproduction, or functional replacement.

discount rate: (Ch. 6) The rate of return required to attract capital to an investment.

discounted cash flow analysis: (Ch. 6) An evaluation of the actual or estimated periodic net income produced by the revenues and expenses in the operation and ultimate resale of an income-producing property.

duty of care: (Ch. 3) An obligation to perform to reasonable standards; to meet the requirements of the assignment.

easement: (Ch. 9) An interest in another's land granting a right to use or control the land (or an area above or below it) for a specific and limited purpose.

ecosystem services: (Ch. 2) The ecological benefits produced by a resource.

ecosystem value: (Ch. 2) The value that ecosystem services produced by plants and other natural resources (such as habitat, clean air, water, and wildlife) contribute to society. Typically a non-market value.

eminent domain: (Ch. 9) The power of a government entity to take private property for public benefit. Also referred to as *expropriation* in some jurisdictions.

ethics: (Ch. 8) Principles that guide conduct and behavior.

existence value: (Ch. 2) A consumer's willingness to pay for the assurance that something remains in existence, even though they may never use or consume it. A type of option value, non-use value, or non-market value.

external limitation: (Ch. 2) A form of depreciation external to the site and outside the control of the property owner that diminishes a plant's value.

extraordinary assumptions: (Ch. 3) Assumptions that presume as fact uncertain information that, if not true, could have a material effect on the appraiser's conclusions.

extrapolated cost: (Ch. 5) An estimate of functional replacement or reproduction cost based on extrapolation.

extrapolation: (Ch. 5) A formulaic method of inferring from the total or unit cost of a small tree what the theoretic installed cost or value of a larger tree might be.

fair market value: (Ch. 2) See *market value*.

forest: (Ch. 9) An ecosystem dominated by extensive tree cover. May be natural or planted. Often synonymous with *woodland*.

form: (Ch. 4) A description of a plant's habit.

functional limitations: (Ch. 2) Defects caused by a flaw in the materials or design of an element.

functional replacement: (Ch. 5) The production of a copy of an existing item that has the same functional utility and is updated to current standards with deficiencies and superadequacies removed.

functional replacement cost: (Ch. 2) The cost to replace a landscape item with an item having equivalent utility.

goodwill: (Ch. 2) An intangible business asset, such as name recognition and reputation, customer and vendor relationships, proprietary operating systems, and quality management. Features that cause the marketplace to pay a premium over real estate market value.

hedonic regression analysis: (Ch. 7) A method of estimating demand or value based on revealed preferences, used to assess the economic values of certain environmental services and their influence on the market price.

heritage tree: (Ch. 9) A tree specimen considered significant due to its size, form, shape, beauty, age, color, rarity, genetic constitution, historic origin, or some other distinctive feature.

highest and best use (HBU): (Ch. 2) The reasonably probable and legal use of a property that is physically possible and financially feasible and results in the highest value. The cornerstone of market value estimation.

hypothetical condition: (Ch. 3) A condition stipulated by the appraiser that is contrary to what is known to be true (e.g., the condition of a damaged plant before the damage occurred, new zoning).

hypsometer: (Ch. 4) A tool for determining tree height by means of triangulation.

investment value: (Ch. 2) A value to a particular owner or investor, as opposed to typical buyers representing the marketplace. Here, the appraiser employs assumptions specific to a particular owner or investor.

i-Tree Eco: (Ch. 2) A suite of public-domain software tools for urban and rural forestry that provide assessment of structure, function, and benefits. Developed by the U.S. Forest Service.

law of diminishing returns: (Ch. 2) A principle asserting that incremental capital or agents of production produce increasing profits or value up to a point, after which added units produce decreasing increments of profit or value in relation to their cost.

limiting conditions: (Ch. 3) Constraints imposed upon appraisals; often associated with items such as the scope of work, the availability of data, access, and use of the report.

liquidation value: (Ch. 2) The most probable price that a property will sell for after being exposed to the marketplace for an abbreviated period of time or to a short list of buyers, and where the seller is usually under extraordinary compulsion to sell.

market value: (Ch. 2) The most probable price, as of a specified date, in cash, in real terms equivalent to cash, or in other precisely revealed terms, for which the specified property rights should sell after reasonable exposure in a competitive market under all conditions requisite to a fair sale, with the buyer and seller each acting prudently, knowledgeably, and for self-interest, and assuming that neither is under undue duress.

multiple regression analysis: (Ch. 7) A form of regression analysis where there are two or more independent variables—for instance, where the independent variables are (X_1) the presence of shade trees and (X_2) property type, and the dependent variable is (Y) increased property value. See *regression analysis*.

non-market valuation: (Ch. 2) A process for estimating the value of resources that are not commonly bought and sold in the marketplace.

non-market value: (Ch. 2) A general term for any value that is not based on the concept of exchange or rooted in actual transactions between buyers and sellers, or cannot be determined from a market price.

parity: (Ch. 5) The state of being equivalent in size or quality, or producing equivalent benefits.

partial loss: (Ch. 4) A situation in which a damaged plant or landscape item cannot be fully restored to its predamage condition but is otherwise expected to produce future benefits.

personal property: (Ch. 2) Tangible property that is not permanently affixed to the real estate, such as nursery stock prior to planting, lumber, equipment, furniture, and money. Also refers to intangible property. See *goodwill.*

physical deterioration: (Ch. 2) A loss in value caused by wear and tear or other physical deterioration. May be curable or incurable.

plant appraisal: (Ch. 2) The act or process of formulating an opinion of a defined value or a defined cost. This may apply to plants, landscape elements, or services.

precise: (Ch. 4) Exact; definite; clearly expressed.

precision: (Ch. 4) A measure of exactness. How fine or precise a quantity, measurement, or estimate is. A number can be precise without being accurate (e.g., an incorrect or misleading number carried out to six significant digits, or a finely measured sample of unrepresentative data).

price: (Ch. 2) The amount of money asked, offered, or paid for property or services.

principle of anticipation: (Ch. 2) A principle asserting that value is created by the expectation of future benefits.

principle of balance: (Ch. 2) A principle asserting that property value is created when contrasting, opposing, or interacting components are in a state of equilibrium.

principle of conformity: (Ch. 2) A principle asserting that property value is created and sustained when the property characteristics conform to the demands of its market as expressed in economic pressures and shared preferences for certain types of structure, amenities, and services.

principle of consistent use: (Ch. 2) A principle asserting that land should not be valued based on one use while improvements on the land are valued based on another use.

principle of contribution: (Ch. 2) A principle asserting that the value of a particular component of a property is measured in terms of its contribution to the value of the whole property, or the amount by which its absence would detract from the value of the whole (attributed to Appraisal Institute 2001).

principle of substitution: (Ch. 2) A principle asserting that when several similar or commensurate commodities, goods, or services are available, the one with the lowest price will attract the greatest demand and widest distribution; value of a replacement property is dictated by the value of an equally desirable substitute property (attributed to Appraisal Institute 2001).

property: (Ch. 2) A term used in appraisal to refer to real estate, real property, or personal property. In plant appraisal, *property* often refers to the real estate of which the item being appraised is a part.

public interest value: (Ch. 2) A general term for values that are recognized by, or accrue to, the general public. Public interest value is often associated with property held or sought by a government agency, and it is derived from intangible ecosystems and recreation values that private markets may not recognize.

punitive damages: (Ch. 9) The amount determined by a court or other entity and levied against a defendant as punishment for wrongful acts and to discourage similar behavior.

qualitative data: (Ch. 4) Data that are usually more descriptive than empirical, often relating to or involving comparisons based on qualities.

quantitative data: (Ch. 4) Data that are readily measured or have a natural order or ranking.

real estate: (Ch. 2) Physical land and appurtenances attached to the land, including the land, things that are naturally part of the land, and things that are attached to the land by people. Native or installed trees and landscaping are real estate (Appraisal Institute 2002).

real estate appraisal: (Ch. 2) The act of estimating the cost or value of real estate.

real property: (Ch. 2) All the interests, benefits, and rights accruing to the owner of real estate, otherwise described as the bundle of rights; simply put, the legal interest in real estate.

reconciliation: (Ch. 3) Part of the appraisal process that involves the resolution of disparate indications of value or cost into a meaningful, defensible conclusion.

regression analysis: (Ch. 7) A mathematical method for quantifying whether there is a statistically significant relationship between one or more independent variables and a particular dependent variable. May be used to predict dbh (Y) from stump diameter (X), or to define how property value (Y) is affected by the presence of trees (X). See *multiple regression analysis.*

replacement cost: (Ch. 2) The cost to replace an improvement with a similar item having equivalent functional utility. Referred to in previous editions of the *Guide* as *cost of cure.*

reproduction cost: (Ch. 2) The cost to replace an improvement with an exact replica. Referred to in previous editions of the *Guide* as *replacement cost.*

right-of-way: (Ch. 9) A legal right to passage or conveyance from one point to another across one or more parcels of land.

salvage value: (Ch. 9) Estimated amount that can be obtained for a component of property after it is no longer useful as part of the whole property (e.g., fuelwood from a dead shade tree, boards from a fence or building).

scarcity: (Ch. 2) The limited supply of an item, often causing it to be worth more in the marketplace.

standard: (Ch. 1) A document established by consensus or government agency that provides rules, guidelines, or characteristics for activities or results.

statutory value: (Ch. 3) A value specified in a statute or law.

stumpage value: (Ch. 2) The value of standing timber to be sold into local timber markets, generally applied to volume (cubic meters, cords, board feet) or weight (tons). A type of market value.

superadequacy: (Ch. 2) An excess in the capacity or quality of a structure or structural component that does not add value or functional utility to an object or property.

supply: (Ch. 2) The quantity of a product or service the market can offer.

timber: (Ch. 9) Standing trees or stumpage available now or in the future for sale as forest products such as fuelwood, pulpwood, sawtimber, or veneer.

trunk formula technique (TFT): (Ch. 5) A technique for developing a cost basis that involves extrapolating the purchase cost of a nursery-grown tree up to the size of the subject tree being valued.

Uniform Standards of Professional Appraisal Practice (USPAP): (Ch. 1) A multipart guide representing generally accepted and recognized standards of appraisal practice in the United States, as developed, interpreted, amended, and published by The Appraisal Foundation.

unit rule: (Ch. 9) A rule stating that the value of the sum of component parts cannot exceed the value of the whole. Different physical elements or components of a tract of land (such as the value of timber and the value of minerals on the same land, irrigated cropland and dry cropland on the same parcel, etc.) are not to be separately valued and added together.

utility: (Ch. 2) The usefulness or benefits provided by an item, giving rise to demand for it.

valuation: (Ch. 2) The act or process of developing an opinion of value. Interchangeable with *appraisal.*

value: (Ch. 2) The monetary worth of a property, good, or service to buyers and sellers at a given point in time. Expectation or present worth of future benefits. Economic value is created by scarcity restricting supply and utility enhancing demand. Not to be confused with *cost* or *price.*

value estimate: (Ch. 2) An assignment result in which the plant appraiser estimates the economic value of a plant or landscape item based on its market supply and demand.

willingness to pay (WTP): (Ch. 2) The idea that an item's value can be demonstrated empirically or inferentially by how much someone is willing to pay to obtain, preserve, or protect it. Can be applied to both market and non-market valuation.

References

American National Standards Institute. 2017. American National Standard for Arboricultural Operations–Safety Requirements (Z133). Champaign, IL: International Society of Arboriculture.

American National Standards Institute. 2014. American Standard for Nursery Stock. ANSI Z60.1. AmericanHort. Columbus, OH.

American National Standards Institute. 2014. ANSI Z60.1-2014 American Standard for Nursery Stock. Washington, DC: American Nursery & Landscape Association.

American National Standards Institute. 2014. ANSI A300–American National Standard for Arboricultural Operations. Part 1. Pruning. Tree Care Industry Association. Londonderry, NH.

American Society of Consulting Arborists. 2013. *Example Reports for Consulting Arborists.* 3rd ed. Rockville, MD.

Anderson, L. and H. Cordell. 1988. Influence of Trees on Residential Property Values in Athens, Georgia (USA): A Survey Based on Actual Sales Prices. *Landscape Urban Planning* 15:153–164.

Angwin, P., D. Cluck, P. Zambino, B. Oblinger, and W. Woodruff. 2012. Hazard Tree Guidelines For Forest Service Facilities and Roads in the Pacific Southwest Region. USDA Forest Service. Forest Health Protection. Pacific Southwest Region. Report # RO-12-01. Redding, CA. http://a123.g.akamai.net/7/123/11558/abc123/forestservic.download.akamai.com/11558/www/nepa/ 99921_FSPLT3_2438910.pdf

The Appraisal Foundation. 2016. *Uniform Standards of Professional Appraisal Practice.* 2016–2017 edition. Washington, DC: The Appraisal Foundation.

The Appraisal Foundation. 2018. *Uniform Standards of Professional Appraisal Practice.* 2018–2019 edition, p. 335. Washington, DC: The Appraisal Foundation.

Appraisal Institute. 1996. *The Appraisal of Real Estate.* 11th ed. Appraisal Institute, Chicago, IL. 820 pp. PIV p. 27.

Appraisal Institute. 2001. *The Appraisal of Real Estate.* 12th ed. Appraisal Institute, Chicago, IL. 759 pp. PIV pp. 649–650.

Appraisal Institute. 2008. *The Appraisal of Real Estate.* 13th ed. Appraisal Institute, Chicago, IL. 742 pp. PIV pp. 31–32.

Appraisal Institute. 2010. *Dictionary of Real Estate Appraisal.* 5th ed. p. 202.

Appraisal Institute. 2013a. *The Appraisal of Real Estate.* 14th ed. Chicago, IL: Appraisal Institute. 847 pp.

Appraisal Institute. 2013b. *Understanding the Appraisal.* Chicago, IL: Appraisal Institute. http://www.appraisalinstitute.org/assets/1/7/understand_appraisal_1109_(1).pdf

Appraisal Institute. 2015. *The Dictionary of Real Estate Appraisal.* 6th ed. Chicago, IL: Appraisal Institute. 527 pp.

Bloch, L. 2000. *Tree Law Cases in the USA.* Bloch Consulting Group. Potomac, MD.

Bonapart, B.K. 2014. *Understanding Tree Law: A handbook for practitioners.* Thomson Reuters/Aspatore Publishing.

Bond, J. 2012. *Urban Tree Health: A Practical and Precise Estimation Method.* Geneva, NY: Urban Forest Analytics, LLC.

Bond, J. 2014. *Best Management Practices: Tree Inventories.* 2nd ed. Champaign, IL: International Society of Arboriculture.

Burns, R. and B. Honkala. 1990. Silvics of North America. USDA Forest Service Agricultural Handbook 654. Washington, DC: USDA Forest Service. http://www.na.fs.fed.us/spfo/pubs/silvics_manual/table_of_contents.htm

Coder, K.D. 2000. Crown Shape Factors & Volume. University of Georgia Cooperative Extension Service. Tree Biomechanics Series Publication FOR00-32. Accessed August 19, 2010. http://warnell.forestry.uga.edu/SERVICE/LIBRARY/for00-032/for00-032.pdf

Costello, L., E. Perry, N. Matheny, J. M. Henry, and P. Geisel. 2003. Abiotic Disorders of Landscape Plants: A Diagnostic Guide. Publication 3240. Berkeley: University of California Agricultural and Natural Resources Communication Services.

Council of Tree & Landscape Appraisers. 1979. *Guide for Establishing Values of Trees and Other Plants.* Revision IV. International Society of Arboriculture. Urbana, IL.

Council of Tree & Landscape Appraisers. 1983. *Guide for Establishing Values of Trees and Other Plants.* Revision V. International Society of Arboriculture. Urbana, IL.

Council of Tree & Landscape Appraisers. 1986. *Manual for Plant Appraisers.* Council of Tree & Landscape Appraisers. Washington, DC.

Council of Tree & Landscape Appraisers. 1988. *Valuation of Landscape Trees, Shrubs, and Other Plants: A guide to the methods and procedures for appraising amenity plants.* 7th ed. International Society of Arboriculture. Champaign, IL.

Council of Tree & Landscape Appraisers. 1992. *Guide for Plant Appraisal.* 8th ed. International Society of Arboriculture. Champaign, IL.

Council of Tree & Landscape Appraisers. 2000. *Guide for Plant Appraisal.* 9th ed. International Society of Arboriculture. Champaign, IL.

Cullen, S. 2005. Tree Appraisal: Chronology of the North American Industry Guidance. *Journal of Arboriculture* 31(4):157–162.

Cullen, S. 2014. Contributory real estate market value of trees: Historical perspective and current competence requirements. *Arboricultural Consultant* 47(4):15–20.

Cullen, D.S. 2015. Measurement Height of Trunk Caliper. *Arboricultural Consultant* 48(1): 11–16.

Curtis, R.O. and D.D. Marshall. 2000. Why Quadratic Mean Diameter? *Western Journal of Applied Forestry* 15:137–139.

Dombrow, J., M. Rodriguez, and C.F. Sirmans. January 2000. The Market Value of Mature Trees in Single-Family Housing Markets. *The Appraisal Journal* 68(1).

Donovan, G. and D. Butry. 2010. Trees in the city: Valuing street trees in Portland, Oregon. *Landscape and Urban Planning* 94:77–83.

Dunster, J. 2014. *Documenting Evidence: Practical Guidance for Arborists.* Dunster & Associates Environmental Consultants Ltd. First Choice Books. Victoria, BC, Canada.

Federal Emergency Management Agency. 2013. Robert T. Stafford Disaster Relief and Emergency Assistance Act, as amended, and Related Authorities as of April 2013. https://www.fema.gov/media-library/assets/documents/15271?fromSearch=fromsearch&id=3564 (accessed December 21, 2016).

Felt, E. 1938. *Our Shade Trees.* New York, NY: Orange Judd Publishing.

Fite, K. and E.T. Smiley. 2016. *Best Management Practices: Managing Trees During Construction.* 2nd ed. Champaign, IL: International Society of Arboriculture.

Harris, R., J. Clark, and N. Matheny. 2004. *Arboriculture – The Integrated Management of Landscape Trees, Shrubs and Vines.* 4th ed. Englewood Cliffs, NJ: Prentice-Hall Inc.

Helms, J. 1998. *The Dictionary of Forestry.* Society of American Foresters. Washington D.C.

Henry, M.S. 1994. The contribution of landscaping to the price of single-family houses: A study of home sales in Greenville, South Carolina. *Journal of Environmental Horticulture* 12(2):65-70.

Husch, B., T.W. Beers, and J.A. Kershaw, Jr. 2002. *Forest Mensuration.* 4th ed. Hoboken, NJ: John Wiley & Sons, Inc.

Interagency Land Acquisition Conference. 2000. Uniform Appraisal Standards for Federal Land Acquisition. B-13, p. 54.

Internal Revenue Service. 2015. *Publication 547 Casualties, Disasters, and Thefts*. Washington, DC.

International Organization for Standardization. 2004. *ISO/IEC Guide 2:2004 Preview Standardization and related activities – General vocabulary.* Geneva, Switzerland: International Organization for Standardization.

Jonnes, J. 2016. *Urban Forests: A Natural History of Trees and People in the American Cityscape.* New York: Viking Press.

Keefer, C. 2004. *A Consultant's Guide to Writing Effective Reports.* 2nd ed. American Society of Consulting Arborists. Rockville, MD.

Komen, J. 2016. Critical Analysis of the ATA Formula. *Arboricultural Consultant* 49(1):12–16.

Lewis, C., editor. 1970. *Shade Tree Evaluation.* Revision III. Urbana, IL: International Shade Tree Conference.

Maeglin, R. 1979. *Increment Cores: How to collect, handle, and use them.* USDA Forest Service. Forest Products Lab. General Technical Report FPL-25. Madison, WI. http://www.treesearch.fs.fed.us/pubs/9801

McPherson, E.G. 2007. Benefit-based tree valuation. *Arboriculture & Urban Forestry* 33(1):1–11.

Merullo, V. and M. Valentine 1992. *Arboriculture and the Law.* Champaign, IL: International Society of Arboriculture.

National Park Service, undated, a. National Register of Historic Places. https://www.nps.gov/nr/ (accessed December 21, 2016).

National Park Service, undated, b. National Register Bulletin: Guidelines for Evaluating and Documenting Rural Historic Landscapes. https://www.nps.gov/nr/publications/bulletins/nrb30/nrb30_8.htm. (accessed December 21, 2016).

Neely, D., editor. 1975. *Guide to the Professional Evaluation of Landscape Trees, Specimen Shrubs, and Evergreens.* Revision III. Champaign, IL: International Society of Arboriculture.

Neely, D., editor. 1979. *Guide for Establishing Values of Trees and Other Plants.* Revision IV. Champaign, IL: International Society of Arboriculture.

Neely, D., editor. 1983. *Guide for Establishing Values of Trees and Other Plants.* 6th ed. Champaign, IL: International Society of Arboriculture.

Neely, D., editor. 1988. *Valuation of Landscape Trees, Shrubs, and other Plants: A Guide to the Methods and Procedures for Appraising Amenity Plants.* 7th ed. Champaign, IL: International Society of Arboriculture.

Nowak, D., D. Crane, and J. Dwyer. 2002. Compensatory Values of Urban Trees in the United States. *Journal of Arboriculture* 28(4):194–199.

Nowak, D., J. Walton, J. Baldwin, and J. Bond. 2015. Simple Street Tree Sampling. *Arboriculture & Urban Forestry* 41(6):346–354.

Passewitz, G. Ohio State University Fact Sheet F-32. See also http://www.treasury.gov/press/releases/reports/nprm%20101905.pdf (p. 83).

Payne, B.R. and S. Strom. 1975. The contribution of trees to the appraised value of unimproved residential land. *Valuation* 22(2):36–45.

Peterson, K. and T. Straka. 2012. Urban Forest and Tree Valuation Using Discounted Cash Flow Analysis: Impact of Economic Components. *Open Journal of Forestry* 2(3):174–181.

Shade Tree Evaluation Committee. 1975. *A Guide to the Professional Evaluation of Landscape Trees, Specimen Shrubs, and Evergreens.* Revision III. International Society of Arboriculture. Urbana, IL.

Smiley, E.T., S. Lilly, and N. Matheny. 2017. *Best Management Practices: Tree Risk Assessment.* 2nd ed. Champaign, IL: International Society of Arboriculture.

Stamen, R. 1997. *California Arboriculture Law.* Randall Stamen. Riverside, CA.

Steigerwaldt, E. and L. Steigerwaldt. 2012. *A Practical Guide to Tree Appraisal.* Steigerwaldt Land Services. Tomahawk, WI.

Swiecki, T. J., and E. A. Bernhardt. 2001. *Guidelines for Developing and Evaluating Tree Ordinances.* Phytosphere Research: Vacaville, CA. Accessed September 14, 2011. http://www.isa-arbor.com/education/ resources/educ_TreeOrdinanceGuidelines.pdf

Urban Tree Growth & Longevity Working Group. *A joint program of the International Society of Arboriculture & the Arboricultural Research & Education Academy.* www.urbantreegrowth.org/

USDA Forest Service. Northern Research Station. Forest Health Inventory Assessment and Monitoring. http://www.nrs.fs.fed.us/inventory_monitoring/

USDA Forest Service. Northern Research Station. i-Tree Manuals & Workbooks. http://www.itreetools.org/resources/manuals.php

USDA Forest Service. Standardized codes for trees species. http://www.nrs.fs.fed.us/tools/ufore/local-resources/downloads/PDF_UDORE_SPECIES_LIST.pdf

USDA Forest Service. 2015. Field Guides for Standard Measurements. Version 7.0. USDA Forest Service. Forest Inventory and Analysis National Program. http://www.fia.fs.fed.us/library/field-guides-methods-proc/)

USDA Forest Service. 2017. i-Tree Eco Ver. 6.0 Field Guide. 57 pp. https://www.itreetools.org/resources/manuals/Ecov6_ManualsGuides/Ecov6_FieldManual.pdf (accessed February 15, 2017)

USDA Forest Service. 2017. i-Tree Eco Ver. 6.0 User's Manual. 97 pp. https://www.itreetools.org/resources/manuals/Ecov6_ManualsGuides/Ecov6_UsersManual.pdf (accessed February 15, 2017)

Vicary, B. 1990. Appraising Premerchantable Timber. *The Canadian Appraiser*. Bk.2/Vol.34, pp. 24–26.

Vicary, B.P. 1986. *The Theory and Practice of Timberland Appraisal*. Thesis in partial fulfillment of PhD in Forestry, University of Maine. 520 pp.

Ware, R., editor. 2005. *The Law of Damages in Wisconsin*. 4th ed. Madison, WI: State Bar of Wisconsin CLE Books. Chapter 18 Real Estate, Par. 3 (Sec. 18.17, Shade and Ornamental Trees). pp. 17–18.

The West Group. 2005. *American Law Reports 3d.* Thomson Reuters.

Index

Pages listed in **bold** include illustrations, in *italics* include tables, and <u>underscored</u> include examples.

Corrigenda

The items in this table represent corrections to this Guide since its first printing in 2018. Revisions that were made between the 10th edition, second printing, and the revised 10th edition (the third printing) are shown in the rows for pages 84 and 86.

Page / Line or #	Original Text	Corrected Text (Minor grammatical errors are not addressed here.)
3 / 10	1977	1975
4 / 30-31	The *Guide* is not an ANSI standard, even though it is produced through a similar consensus-driven process.	The *Guide* is not an ANSI standard, even though it is produced through a consensus-driven process.
5 / 8-9	...marketplace. The *Guide* meets this standard. It has a long history of...	...marketplace. The *Guide* has a long history of...
9 / 25	...it represents market value.	...it may represent market value.
10 / 2 and throughout	...inferred from software application systems like **i-Tree Eco**...	...inferred from tree management software applications like **i-Tree Eco**...
11/ 6	**Cost Estimates Versus Value Estimates**	**Cost Estimates Versus Market Value Estimates**
11 / 13	Where market value is sought, evidence of WTP derives from transactions.	---
22 / 13	• Contractual value.	---
22 / 19	*Uniform Standards of Appraisal Practice* (USPAP)	*Uniform Standards of Professional Appraisal Practice* (USPAP)

28 / 30	No reconciliation was needed.	No reconciliation was necessary.
30 / 12, 14	25 feet (7.6 m) 3.5 feet (1.5 m)	25 feet (7.62 m) 3.5 feet (1.07 m)
35 / footnote	Palm trunk height is measured from grade to the base of the newest, youngest leaf (also known as the spear leaf).	Trunk height is measured from the ground line, which should be at or near the top of the root zone to the base of the heart leaf (ANSI Z60-2014).
37 / 9-22	The trunk measurement of a leaning tree on level ground should be made 4.5 feet from the ground on the compression or underside of the trunk. Measurement should be perpendicular to the trunk (Figure 4.3b).	The trunk measurement of a leaning tree on level ground should be made 4.5 feet from the ground on the compression or underside of the trunk. Measurement of the trunk diameter should be perpendicular to the trunk (smallest diameter across the trunk) (Figure 4.3b).
37 /11-12	The trunk measurement of a leaning tree on a slope should be made 4.5 feet from the ground on the high side of the trunk. Measurement should be perpendicular to the trunk (Figure 4.3c).	The trunk measurement of a leaning tree on a slope should be made 4.5 feet from the ground on the high side of the slope. Measurement of the trunk diameter should be perpendicular to the trunk (smallest diameter across the trunk) (Figure 4.3c).
38 / 4-7	**Trunk with more than one stem originating at or near ground level.** If all the stems arise within 3 feet (1 m) of the ground and each stem contributes proportionately to the crown, measure the diameter of each stem at 4.5 feet (1.37 m) (Figure 4.3g). Alternatively, measure the trunk diameter of a comparable single-stem tree of similar height and crown spread and apply that measurement.	**Trunk with more than one stem originating at or near ground level.** If all the stems arise from within 3 feet (1 m) of the ground, and each stem contributes equally to the canopy, then determine the sum of the cross sectional areas of each stem measured at 4.5 feet (1.37 m) above grade (see figure 4.3g). Different stem configurations may require measuring at other heights or locations to more accurately represent the size of a stem (see figures 4.3 e-k).

39 / 2-3	3 to 5 feet (1 to 1.5 m)	3 to 5 feet (1 to 1.07 m)
44 / Table 4.1	Excellent [Percent Rating] 100%	Excellent [Percent Rating] 81% to 100%
49 / Table 4.3	**Component Rating Weighting Product** Health 1.00 0.15 0.15 Structure 0.60 0.70 0.42 Form 0.40 0.15 0.06 **Weighted average condition rating** (sum of product) **0.63** **Note:** Weighting factors must add up to 1.00 or 100%.	**Component Rating Weighting Product** Health 1.00 0.15 0.15 Structure 0.60 0.70 0.42 Form 0.40 0.15 0.06 Sum 2.0 1.0 0.63 **Weighted average condition rating** (sum of product/sum of ratings) **0.63 ÷ 1** **Weighted average condition rating 0.63**
50 / 5-13	A third approach employs a weighted average of the three components (Table 4.3; Figure 4.10). This process involves four steps. First, health, structure, and form are evaluated. Second, the appraiser considers whether one of these components is more important than any other, and, if so, applies a weighting factor. Third, the ratings of health, structure, and form are multiplied by the weighting factor. Fourth, the product of the rating and weighting are totaled. For example, the Deodar cedar in Figure 4.10 was assessed with a health rating of 1.00 (100%), a structure rating of 0.60 (60%), and a form rating of 0.40 (40%). In the second step…	A third approach employs a weighted average of the three components (Table 4.3; Figure 4.10). This process involves four steps. First, health, structure, and form are evaluated in decimal form and the results added together. Second, the appraiser considers whether one of these components is more important than any other, and, if so, applies a weighting factor. Third, the ratings of health, structure, and form are multiplied by the weighting factor. Fourth, the product of the rating and weighting are added together and divided by the sum of the original ratings. For example, the Deodar cedar in Figure 4.10 was assessed with a health rating of 1.00 (100%), a structure rating of 0.60 (60%), and a form rating of 0.40 (40%). In the second step…

51 / 1-3	Fourth, the result was totaled to calculate the weighted average of 63%, equivalent to good condition (as suggested in Table 4.1).	Fourth, the product of each component is added together and divided by the sum of the original (0.15 + 0.42 + 0.06) / 1.0. In this example, the result was a weighted average of 63%, equivalent to poor condition (Table 4.1).
54 / Figure 5.1	**Original text** Functional Reproduction Replacement	**Replacement text** Reproduction Functional Replacement
57 / 27-30	Estimates of tree value may be greatly out of proportion to the value of the land and other property improvements, or to what people would actually pay for a replacement tree.	Cost estimates may be greatly out of proportion to the value of the land and other property improvements, or to what people would actually pay for a replacement tree.
57 / 68-69	To apply the TFT using trunk diameter, compute the cross- sectional area of the subject plant then multiply it by the unit price.	To apply the TFT using trunk diameter, compute the cross-sectional area of the subject plant then multiply it by the unit price (see Appendix 2).
58 / 15-16	...important than overall tree size. After all, tree diameter is simply a proxy for tree size. In most cases, tree diameter in and of itself confers little in the way of direct benefits.	...important than overall trunk diameter.

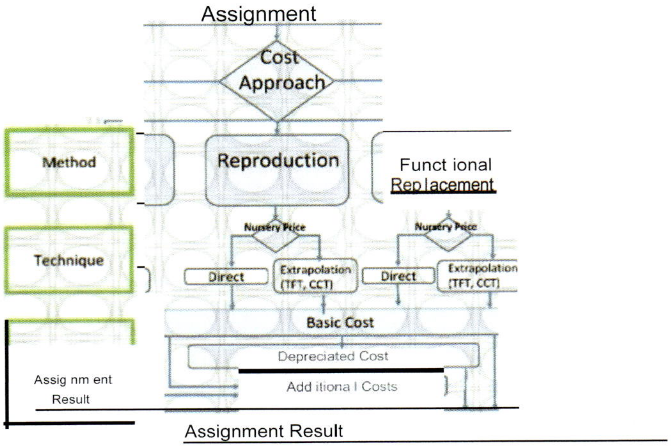

59-60 / 36 1-3	(present cost, *PC*)	(present installed cost, *PC*)
64 / 23	water use limitations, restrictions on irrigation;	water use limitations, restrictions on irrigation; competing infrastructure (utilities);
65 / 11-13	Appraisers may find that some features fit into more than one depreciation category. For example, overhead electrical wires are a functional limitation because they are over the property, but were a tree has been topped because of the powerline, the appraiser may depreciate for both condition (form) and functional limitations that will limit future height growth. The appraiser should not also depreciate for the decision to install the powerlines over the property that was out of the control of the property owner because the physical structure (powerline) is already in place.	Appraisers may find that some features fit into more than one depreciation category. For example, overhead electrical wires can be either a functional limitation or an external limitation. In this case, the appraiser should depreciate in only one category.
65 / 21-2	...prepared by contractors or other professionals.	...prepared by contractors, other professionals, or the appraiser, if qualified and not conflicted.
67-68 & 70-72 / 25, 20, 7, 24, & 13	total (accrued) depreciation	total depreciation
71 / 4-7	3. Weighted average: 47% a) Weighting: structure, 0.40; health, 0.30; form 0.30 b) Weighted average: (50% × 0.40) + (50% × 0.30) + (30% × 0.30) = 20% 20% +15% + 9% = 44%	3. Weighted average: 34% a) Weighting: structure, 0.40; health, 0.30; form 0.30 b) Weighted average: (50% × 0.40) + (50% × 0.30) + (30% × 0.30) = 20% +15% + 9% = 44% ÷ 1 = 44%
71 / 36-38	3. Weighted average: 82% a)Weighting: structure, 0.40; health, 0.40; form, 0.20 b) Weighted average: (90% × 0.40) + (70% × 0.40) + (90% × 0.20) = 36% +28% + 18% = 82%	3. Weighted average: 35% a)Weighting: structure, 0.80; health, 0.10; form, 0.10 b) Weighted average: (90% × 0.80) + (70% × 0.10) + (90% × 0.10) = 72% + 7% + 9% = 88% ÷ 1.0 = 88%

74 / 5	...benefit was to screen Ms. Peabody's home.	...benefit was to screen Mr. Butler's view of Ms. Peabody's home.
75 / 39	Installation cost. 10 trees @ $10.	Installation cost. 10 trees @ $100.
77 / 22-24	The principle of substitution might otherwise argue for using the lowest estimate, of $40, but in this case, the appraiser selects a higher number reflecting intangible benefits of superior tree quality and service.	The principle of substitution might otherwise argue for using the lowest estimate (Nursery 1, $40.74), but in this case the appraiser selects a higher estimate (Nursey 3, $44.56) because of its superior tree quality and reputation for excellent customer service.
79 /12	$42,316	$8,458
79 / 17-18	$\$44/\text{in}^2$	$\$44.56/\text{in}^2$
80 / 6	20-foot dbh (2.5-m)	20-inch dbh (0.58 cm)
83 / 28-30	...computed as follows: Year 1 = $400 (o discount for Year 1) Year 2 = $400 ÷1.05	...computer as follows: Year 1 = $400 (0 discount for Year 1) Year 2 = $\$400 \div 1.05^1$
84 / 6 through end of page		*Except for the first lines, from* Client name *through* Address, *this form has been revised in its entirety. Specifics of the revision can be found on the product page for the* Guide for Plant Appraisal *on ISA's online store.*
85 / #2	2. Cross-sectional area (line $1)^2 \times 0.7854$ =	2. Cross-sectional area (line $1)^2 \times 0.7854$
85 / #7	7. Cross-sectional area (line $6)^2 \times 0.7854$ =	7. Cross-sectional area (line $6)^2 \times 0.7854$
85 / #11	11. Depreciated reproduction cost (line 3 × line 4 × line 5 × line 10)	11. Depreciated reproduction cost [‡] (line 10 × line 3 × line 4 × line 5)
85 / footnote	*dbh and growth rate may be replaced with plant area, volume, or height as appropriate.	*diameter and cross-sectional area may be replaced with plant area, volume, or height as appropriate

86 / #1–#3	1. Trunk diameter* (D) _______ @ _______ 2. Cross-sectional area (line 1)^2 × 0.7854 _________ in^2 3. Condition rating _________ %	1. Trunk diameter* (D) _________ @ _________ 2. Condition rating _________ %
86 / 19 through end of page		*From* Species *through the end of the page, this form has been revised in its entirety. Specifics of the revision can be found on the product page for the* Guide for Plant Appraisal *on ISA's online store.*
87 / #11	11. Basic compounded cost (line 8 [1 + line 10] $^{\text{line 9}}$)	11. Basic compounded cost (line 8 × [1 + line 10]$^{\text{line 9}}$)
87 / #12	12. Depreciated compounded cost (line 3 × line 4 × line 5 × line 11)	12. Depreciated compounded cost ‡ (line 11 × line 2 × line 3 × line 4)
88 / #12	12. Depreciated compounded cost (line 3 × line 4 × line 5 × line 12)‡	12. Depreciated compounded cost ‡ (line 11 × line 2 × line 3 × line 4)
88 / footnote	*dbh and growth rate may be replaced with plant area, volume, or height as appropriate. **the age and diameter growth of the subject tree are not necessarily relevant. Its size (dbh, volume, and/or height) is relevant. Years to parity should reflect the appraise r's best estimate of the time for a healthy specimen to grow to the same basic size as the subject tree.	*diameter and cross-sectional area may be replaced with plant area, volume, or height as appropriate. **The age and cross-sectional area of the subject tree are not necessarily relevant. Its size (diameter, volume, and/or height) is relevant. Years to parity should reflect the appraiser's best estimate of the time for a healthy specimen to reach a size where it provides equal utility or benefits.
98 / 2-3	Highest and best use is foundational for estimating market value…	Highest and best use should be considered a function of the appraisal problem…

98 / 17-19	The price paid for plants at a nursery or for landscape services is the market value of those goods and services. It is set based on supply, demand, and other factors. Estimating the cost of these goods and services is an application of the SCA, but that is not the emphasis in this chapter.	---
101 / 1	Overall, these studies show…	These studies show…
101 / 6-7	SOURCES FOR PROPERTY VALUE Sources for property value include:	SOURCES FOR MARKET VALUE OF PROPERTY Sources for market value of property include:
112 / 1-4	While the plant appraisal profession may not be highly developed…,	--
126 / 21	Other terms applied to trees are *ancient*, *veteran*, *landmark*, *legacy*, and…	Other terms applied to trees are *ancient*, *veteran*, *landmark*, *legacy*, *specimen tree*, and…
127 / 5	At the global level, the United Nations Education, Scientific, and Cultural Organization (UNESCO) designates heritage sites (e.g., Yellowstone National Park).	At the global level, the United Nations Educational, Scientific, and Cultural Organization (UNESCO) designates World Heritage Sites (e.g., Yellowstone National Park, Mammoth Cave National Park, etc.).

128 / 5-7	The limit of insurance (liability) of this coverage for all trees, shrubs, plants, and lawns may not exceed 5% of the limit of liability that applies to the dwelling, or more than $500 for any one tree, shrub, or plant.	There are limits to insurance (liability) for all trees, shrubs, plants, and lawns.
131 / 17	...casualty loss purposes.	...casualty loss purposes. If you encounter a situation that involves the tax code, consult a tax professional.
136 / 28-29	*Betula nigra* Heritage™	*Betula nigra* 'Cully' The (trademarked) common name is Heritage™ River Birch.
136 / 35-39	*Betula nigra* Dura-Heat™ (Actual cultivar name is *Betula nigra* 'Cully')	*Betula nigra* The (trademarked) common name is Dura-Heat™ River Birch.
142 / 9-10	Only basic statistics, commonly employed in forestry, are presented.	---
142 / 12	...appraiser should enlist the services of a professional forester skill in forest inventory...	...should follow industry-standard forest inventory sampling and design.
143 / Table A3.1*		*Significant digits throughout have been updated for consistent accuracy and precision.
151 / 24	Shady Grove Nursery	Shady Creek Nursery
158 / 36	**form:** (Ch. 4) A description of a plant' s habitat.	**form:** (Ch. 4) A description of a plant's habit.
161 / 19-22	**trunk formula technique (TFT)**: (Ch. 5) A technique for developing a cost basis that involves extrapolating the acquisition cost of a nursery-grown tree up to the size of the subject tree being valued.	**trunk formula technique (TFT):** (Ch. 5) A technique for developing a cost basis that involves extrapolating the purchase cost of a nursery-grown tree up to the size of the subject tree being valued.

161 / 23-24	**Uniform Standards of Professional Practice (USPAP)**	**Uniform Standards of Professional Appraisal Practice (USPAP)**
170 / 20	unit rule, 129	unit rule, 123, 129-130

Conversion Chart (Metric to Imperial)

Metric unit	Multiply by	To obtain English unit
Length		
millimeters (mm)	0.04	inches (in)
centimeters (cm)	0.4	inches (in)
meters (m)	3.3	feet (ft)
meters (m)	1.1	yards (yd)
kilometers (km)	0.6	miles (mi)
Area		
square centimeters (cm^2)	0.16	square inches (in^2)
square meters (m^2)	1.2	square yards (yd^2)
square meters (m^2)	10.8	square feet (ft^2)
square kilometers (km^2)	0.4	square miles (mi^2)
hectares (ha)	2.5	acres (ac)
Mass (weight)		
grams (g)	0.035	ounces (oz)
kilograms (kg)	2.2	pounds (lb)
metric tonnes (t)	1.1	short tons
Volume		
milliliters (mL)	0.03	fluid ounces (fl oz)
milliliters (mL)	0.06	cubic inches (in^3)
liters (L)	2.1	pints (pt)
liters (L)	1.06	quarts (qt)
liters (L)	0.26	gallons (gal)
cubic meters (m^3)	35	cubic feet (ft^3)
cubic meters (m^3)	1.3	cubic yards (yd^3)
Temperature (exact)		
degrees Celsius (°C)	multiply by 9/5, then add 32	degrees Fahrenheit (°F)